Trends in

BIOLOGICAL ANTHROPOLOGY

Volume 1

edited by

Karina Gerdau-Radonić
Kathleen McSweeney

Proceedings of the British Association for Biological Anthropology and Osteoarchaeology 13th and 14th Annual Conferences in Edinburgh (2nd–4th September 2011) and Bournemouth (14th–16th September 2012)

Monograph Series Volume 1

Volume editors: Karina Gerdau-Radonić (Bournemouth University) and Kathleen McSweeney (University of Edinburgh) with contributions by Amanda H. Korstjens, Holger Schutkowski and Martin J. Smith

Monograph Series editor: Tina Jakob (Durham University)

Oxbow Books
Oxford & Philadelphia

Published in the United Kingdom in 2015 by
OXBOW BOOKS
10 Hythe Bridge Street, Oxford OX1 2EW

and in the United States by
OXBOW BOOKS
908 Darby Road, Havertown, PA 19083

Paperback Edition: ISBN 978-1-78297-836-7
Digital Edition: ISBN 978-1-78297-837-4

A CIP record for this book is available from the British Library

Printed and bound in the United Kingdom by Hobbs the Printers Ltd, Totton, Hampshire

For a complete list of Oxbow titles, please contact:

UNITED KINGDOM
Oxbow Books
Telephone (01865) 241249, Fax (01865) 794449
Email: oxbow@oxbowbooks.com
www.oxbowbooks.com

UNITED STATES OF AMERICA
Oxbow Books
Telephone (800) 791-9354, Fax (610) 853-9146
Email: queries@casemateacademic.com
www.casemateacademic.com/oxbow

Oxbow Books is part of the Casemate Group

Front cover: Official logo of the British Association for Biological Anthropology and Osteoarchaeology

Table of Contents

Contributors

Vesna Bikić
Institute of Archaeology
Kneza Mihaila 35/IV
11000 Belgrade
Serbia
Email: vesna.bikic@gmail.com

Jo Buckberry
Archaeological Sciences
University of Bradford
Bradford
BD7 1DP
United Kingdom
Email: J.Buckberry@bradford.ac.uk

Dominique Castex
UMR 5199 PACEA-A3P
Bâtiment B8
Université Bordeaux 1
Avenue des Facultés
33405 TALENCE CEDEX
France

Brenna R. Hassett
Natural History Museum London
Cromwell Road
London
SW7 5BD
United Kingdom
Email: b.hassett@nhm.ac.uk

Malcolm Lillie
Department of Geography, Environment and Earth Sciences
University of Hull
Hull
HU6 7RX
United Kingdom
Email: M.C.Lillie@hull.ac.uk

Gabriele A. Macho
Research Laboratory for Archaeology and the History of Art (RLAHA)
Dyson Perrins Building
South Parks Road
Oxford
OX1 3QY
United Kingdom
Email: gabriele.macho@rlaha.ox.ac.uk

Diana Mahoney Swales
Department of Archaeology
University of Sheffield
Northgate House
West Street
Sheffield
South Yorkshire
S1 4ET
United Kingdom

Simon Mays
Archaeological Science
English Heritage
Portsmouth
PO4 9LD
United Kingdom

Iona McCleery
School of History
University of Leeds
Leeds
West Yorkshire
LS2 9JT
United Kingdom
Email: I.McCleery@Leeds.ac.uk

Nataša Miladinović-Radmilović
Institute of Archaeology
Kneza Mihaila 35/IV
11000 Belgrade
Serbia
Email: miladinovic.radmilovic@gmail.com

Alexey G. Nikitin
Biology Department
Grand Valley State University
Allendale
MI 49401
USA
Email: nikitin@gvsu.edu

Pia Nystrom
Department of Archaeology
University of Sheffield
Northgate House
West Street
Sheffield
South Yorkshire
S1 4ET
United Kingdom
Email: p.nystrom@sheffield.ac.uk

Alan Ogden
Archaeological Sciences
University of Bradford
Bradford
BD7 1DP
United Kingdom
Email: A.R.Ogden@Bradford.ac.uk

Inna Potekhina
Department of Bioarchaeology
Institute of Archaeology
Ukrainian Academy of Sciences
12 Heroyiv Stalingrade Avenue
04210 Kiev
Ukraine
Email: potekhina@hotmail.com

Géraldine Sachau-Carcel
UMR 5607 AUSONIUS
Université Bordeaux 3
Maison de l'Archéologie
Esplanade des Antilles
33607 PESSAC CEDEX
France
Email: g.sachau@wanadoo.fr

Vicky Shearman
Wakefield Museums
Wakefield Council
Wakefield One
Burton Street
Wakefield
WF1 2EB
United Kingdom
Email: vshearman@wakefield.gov.uk

Mykhailo P. Sokhatsky
Verteba Excavation Director
Borschiv Regional Museum
Ministry of Culture and Arts of Ukraine
Shevchenka St. 9 Borschiv
Ukraine
Email: sokhal@rambler.ru

Douglas H. Ubelaker
Department of Anthropology
National Museum of Natural History
MRC 112
Smithsonian Institution
Washington, DC 20560
USA
Email: ubelaked@si.edu

Robert Vergnieux
UPS SHS 3D no. 3551
Archéovision
Université de Bordeaux 3
Esplanade des Antilles
33607 PESSAC CEDEX
France

Stefanie Vincent
Archaeological Science,
English Heritage
Portsmouth
PO4 9LD
United Kingdom
Email: stefanie_ _ vincent@hotmail.com

A. Gaynor Western
Ossafreelance
8, North Street
Old Market
Wisbech
Cambridgeshire
PE13 1NP
United Kingdom
Email: info@ossafreelance.co.uk

Marlo Willows
University of Edinburgh
School of History, Classics, and Archaeology
Doorway 4, Teviot Place
Edinburgh
EH8 9AG
United Kingdom
Email: m.willows@sms.ed.ac.uk

The Trends in Biological Anthropology Series

Piers D. Mitchell
President of the British Association for Biological Anthropology and Osteoarchaeology

The first annual conference of the British Association for Biological Anthropology and Osteoarchaeology (BABAO) was held in 1999. For many years papers from the conference were published as a conference proceedings volume. However, the association has grown and matured over time so that we are now in a position to establish a new book series. The *Trends in Biological Anthropology* series will gradually build up a body of quality research articles across all fields of our specialty. This ranges from burial archaeology to modern human genetic variation, from disease in the past to modern forensics, and from human evolution to modern primatology. Articles will include those presented at our national conference, as well as others that may be submitted for consideration. Those hosting the annual conference each year will edit the relevant volume, with support from Tina Jakob the series editor. All articles undergo double peer review, and will only be accepted if they meet a standard we would expect in a peer-reviewed journal.

The content of each volume is likely to reflect the particular interests of the university that hosted the conference, as special themed sessions are always held each year. Well-known researchers from across the country are invited to present at such sessions, and often keynote speakers are invited from abroad in order to stimulate debate and add an international flavour to the session. Articles from these themed sessions will ensure that each volume has its own distinctive character.

We on the executive committee of BABAO are confident that the *Trends in Biological Anthropology* series will flourish and provide a great output for research in the field. We do hope you enjoy this volume and its successors in the series, and that you will consider publishing your own work in future volumes.

Introduction

This volume brings together some of the papers presented during the British Association for Biological Anthropology and Osteoarchaeology annual meetings, 2011, held at the University of Edinburgh, and 2012, held at Bournemouth University. The Edinburgh conference organisers included Kathleen McSweeney, Elena Kranioti, Marlo Willows and Dawn Gooney and the sessions revolved around palaeopathology, scientific advances in osteology, and forensic anthropology, subject areas that are representative of individual fields of interest in the human osteology team at the University of Edinburgh. Bournemouth University's Biological Anthropology team – Martin Smith, Karina Gerdau-Radonić, Holger Schutkowski, Elizabeth Craig-Atkins and Amanda Korstjens – organised the 2012 meeting which offered sessions in primatology and human evolution, interpreting and analysing trauma, interpreting and analysing funerary deposits and the regular open session. The range of papers presented at the meetings and those finally included in this volume are representative of the present breadth of research in Biological Anthropology and Osteoarchaeology within Britain and abroad.

We decided to organise the volume according to the themes that emerged from the papers submitted. The volume opens with studies focusing on extant non-human primates. Gabriele Macho's *Can Extant Primates Serve as Models to Determine the Dietary Ecology of Hominins? The Case of Paranthropines* covers the study of modern baboons (*Papio cynocephalus*) as proxies to understand extinct hominin species' diets and hence our own evolutionary history. Diana Mahoney Swales and Pia Nystrom's *Recording Primate Spinal Degenerative Joint Disease Using a Standardised Approach* focuses on the use of human standards to analyse and interpret skeletal degenerative joint disease (SDJD) on the skeletal remains of extant primates, in particular within the superfamily Cercopithecoidea. Their study looks at the impact of locomotion and body mass on the development of SDJD.

These chapters on extant primates are followed by two on methods used in the fields of Biological Anthropology and Osteoarchaeology. Brenna Hassett's *Enamel Hypoplasia in Post-Medieval London: A Reassessment of the Evidence for Childhood Health* is a study on the methods used to record Linear Enamel Hypoplasia (LEH) and how the choice of method can affect the prevalence recorded in different populations and hence the conclusions drawn from these types of studies. Géraldine Sachau-Carcel, Dominique Castex, and Robert Vergnieux's *Archaeoanthropology: How to Construct a Picture of the Past?* focuses on the use of three-dimensional modelling to generate pictures of the content of collective graves. These images can then function as a means to understand the use and history of these structures.

These two papers are followed by three others that focus on Palaeopathology and Trauma. Marlo Willows' *Palaeopathology of the Isle of May* presents the palaeo-pathological analysis of a skeletal collection from the Isle of May, off the coast of Scotland, dating between the 5th and 16th century AD. By comparing this collection to one from medieval Scotland, Willows tries to ascertain whether the palaeopathological data is evidence that the Isle of May benefitted from a healing tradition during the medieval period, as historical records and legends would have it. Malcolm Lillie, Inna Potekhina, Alexey G. Nikitin and Mykhailo P. Sokhatsky's *First Evidence for Interpersonal Violence in Ukraine's Trypillian Farming Culture: Individual 3 from Verteba Cave, Bilche Zolote* is a case study of a cranium found at Verteba Cave, western Ukraine, as a means to understand inter-personal interactions and burial ritual in that part of Ukraine during the Trypillian culture. This section is closed by Nataša Miladinović-Radmilović and Vesna Bikić's *Beheading at the Dawn of the Modern Age: The Execution of Noblemen during Austro-Ottoman Battles for Belgrade in the Late 17th Century*, the study of a series of skulls found at a deposit site outside the fortress walls of Belgrade, Serbia, displaying evidence for beheading. Through the use of osteological and archaeological data, this study contributes to the history of Belgrade.

Two papers focus on the History of Medicine and Biological Anthropology. Gaynor Western's *The Remains of a Humanitarian Legacy: Bioarchaeological Reflections of the Anatomized Human Skeletal Assemblage from the Worcester Royal Infirmary* presents the results of the excavation and analysis of a deposit containing disarticulated human remains at the Worcester Royal Infirmary. Western's paper, though presenting osteological and archaeological evidence, is a contribution to our understanding of the History of Medicine and how the study of human anatomy progressed and developed. Stefanie Vincent and Simon Mays' *Thomas Henry Huxley (AD 1825–1895): Pioneer of Forensic Anthropology* revisits Huxley's analysis and report on the skeletal remains of a member of Sir John Franklin's 1845 expedition to the Arctic and highlights one of the earliest attempts at identifying a recently deceased individual through the analysis of his or her skeletal remains as do forensic anthropologists today.

Finally, we close the volume with two papers that illustrate how the study of human remains whether modern or past can contribute to the modern world. Douglas Ubelaker's *The Concept of Perimortem in Forensic Science* defines 'perimortem', particularly within a forensic anthropology context. He emphasises the importance of this definition because though a forensic anthropologist will not determine the cause and manner of death, he or she will interpret the timing of the injury, which may in turn contribute to the forensic pathologist's determination of cause and manner of death. Jo Buckberry, Alan Ogden, Vicky Shearman and Iona McCleery's *You Are What You Ate: Using Bioarchaeology to Promote Healthy Eating* presents a collaborative effort between historians, archaeologists, museum officers, medieval re-enactors and food scientists to encourage healthy eating among present day Britons by presenting the ill effects of certain dietary habits on the human skeleton.

To conclude, we thank our colleagues Martin Smith, Holger Schutkowski, and Amanda Korstjens for their editorial input and contributions, the team of reviewers, Laszlo Bartosiewicz, Laura Bassell, Sean Beer, Anthea Boylston, Elizabeth Craig-Atkins, Roxana Ferllini, Louise Loe, Piers Mitchell, Robert Paine, Catriona Pickard, Katherine Robson-Brown, Todd Rae, Norman Sauer, Rick Schulting, and John Stewart, the British Association for Biological Anthropology and Osteoarchaeology's committee, Tina Jakob, the series editor, and last but not least, the contributors for their patience and hard work.

Commemoration

Given the subsequent sad death of our friend and colleague, Professor Donald Ortner, in 2013, we were extremely privileged that he was able to open the 2011 Edinburgh conference with one of his ever-stimulating keynote speeches. He will be sincerely missed.

Karina Gerdau-Radonić
(Bournemouth University)
Kathleen McSweeney
(University of Edinburgh)

1. Can Extant Primates Serve as Models to Determine the Dietary Ecology of Hominins? The Case of Paranthropines

Gabriele A. Macho

The dietary ecology of early hominins, particularly East and South African Paranthropus, *remains poorly understood. Here I argue that an integrative approach that combines current knowledge on isotope composition, microwear textures, dental morphology and comparative studies on the extant baboon* Papio cynocephalus *has the potential to shed light on the possible diet(s) of* Paranthropus boisei *and* P. robustus.

Baboons eat a variety of C_4 foods, which differ considerably in nutritional value and material properties. East and South African paranthropines apparently spent longer periods of time feeding on similar C_4 foods; their morphology suggests that they exploited opposite ends of the C_4 plant food niche spectrum that is utilised by baboons. Paranthropus boisei *consumed predominantly hard brittle foods, while* P. robustus *fed on hard tough resources. Because of the high nutrional value of some C_4 foods, a shift in dietary preferences from C_3 to C_4 sources need not have been accompanied by an extension of total feeding time. To what extent differences in food selection and time spent feeding on C_4 foods between* P. boisei *and* P. robustus *were due to habitat differences between East and South Africa, or constitute true species preferences, needs to be investigated further.*

Keywords: *Paranthropus boisei; Paranthropus robustus;* Hominin dietary ecology; C_4 plant niche; *Papio cynocephalus*

1. Introduction

The dietary adaptations of extinct hominins remain the topic of intense research. This is unsurprising as diet underpins all aspects of an animal's biology and a shift in diet facilitated the marked life history changes seen in our lineage, including the 3-fold increase in brain size (Leonard and Robertson, 1997). Determining the dietary niche(s) of extinct hominins is not trivial however and is hampered by a number of factors. The hominin masticatory apparatus differs greatly from that of extant primates and may be phylogenetically constrained, while the fragmentary nature of the hominin fossil record makes it impossible to carry out in-depth functional analyses. Therefore, palaeoanthropological studies have tended to focus on information that can be gleaned from dental remains, particularly enamel thickness (e.g. Rabenold and Pearson, 2011), as well as microwear textures and isotope composition. The latter two approaches are regarded particularly useful in deciphering the dietary ecologies of hominins, as microwear and isotope composition directly reflect what an animal ate during life (Grine *et al.*, 2012). Despite considerable research efforts in these areas however, the diets of hominins remain poorly understood. As a case in point, *Paranthropus boisei* from East Africa is assumed to have subsisted almost exclusively on a low-quality diet of grasses and sedges (up to 91%), as inferred from the isotope composition of their hard tissue (Lee-Thorp, 2011; Cerling *et al.*, 2013). Yet, such interpretations are inconsistent with *P. boisei* morphology and its energetic requirements predicted from body mass and brain size; this interpretation is also in conflict with the species´microwear textures, which resemble those of a soft fruit consumer (Ungar *et al.*, 2008). *Paranthropus robustus* from South Africa, in contrast, is similar to gracile australopith with regard to isotope composition, while overall morphological and cladistic considerations predict its feeding ecology to have been comparable to that of *P. boisei*. Clearly, not all predictions can be equally valid.

Of course, hominins may have been truly unique and their ecologies never be known. While this is possible, it is improbable. Food sources do not vary so dramatically across Africa, nor between now and the past. I therefore propose a bottom-up approach that (a) explores the variety of foods available to primates with similar physiologies, specifically baboons, and (b) assesses the capabilities of the hominin masticatory apparatus to break down these various food types. It is imperative to identify what an animal would *not* have been able to eat on the basis of its morphology before exploring its potential niche, i.e., what it *could have* eaten (availability). Another important factor for determining the diet of an animal is energetics: Could the inferred diet have provided the hominin with sufficient energy and nutrients while, at the same time, being low in toxins (Altmann, 2009)? Could the animal have harvested the proposed foods within its time budget constraints? To answer these questions a multi-faceted approach is necessary that combines information derived from morphology, behavioural ecology of extant primates and modelling. Here I outline the fundamentals of such an approach and explore whether the results obtained from each strand could be brought together to arrive at a coherent inference about the dietary niche(s) of *P. boisei* and *P. robustus*. Papionins serve as a modern template.

Papio and *Theropithecus* have long been considered good analogs for an appraisal of the dietary radiation of hominins vis-à-vis the ecological drivers underlying it (Jolly, 1970, 2001; Elton, 2006). Together with suids, hominins and baboons have the same basic physiology and share(d) the same ecological and, presumably, dietary niche (Hatley and Kappelman, 1980), i.e., they exploit(ed) Underground Storage Organs (USOs). Baboons, modern humans and pigs are the only large-bodied mammals that can, and habitually do, extract below-surface foods. Together with many sedges and grasses USOs mostly follow the C_4 photosynthetic pathway. Rather than being of low nutritional value however, USOs tend to be nutrient-rich and could therefore have constituted a valuable source for large-bodied large-brained hominins. In fact, it has been suggested that USOs may have played *the* key role in the evolution of modern human life histories and sociality, including allocare (O´Connell *et al.*, 1999, 2002). Comprehensive nutrional analyses of C_4 foods are unfortunately still wanting, in part because C_4 foods are generally considered nutritionally unimportant and/or, as in the case of USOs, because they require substantial extraoral preparation, i.e., pounding or cooking, before they are suitable for human (hominin?) consumption (Carmody *et al.*, 2011) and have therefore been thought to have played a minor role in the earliest stages of human evolution (but see Dominy *et al.*, 2008). Fortunately, empirical data on the feeding ecology of yearling baboons from the Amboseli National Park, Kenya, are available and are sufficiently detailed (Altmann, 1998) to inform which C_4 foods are commonly eaten by baboons and their nutritional value. This allows to assess (a) whether early hominins could have subsisted on a predominantly C_4 diet, (b) whether they could have obtained these foods within their daily time budget, and (c) whether such a diet is consistent with hominin dento-cranial morphologies.

2. Background

2.1. The Palaeontological Background

With its hyper-masticatory apparatus and thick-enamelled teeth *P. boisei* has originally been interpreted as a hard object feeder (Tobias, 1967), but microwear texture analyses seem to indicate that it fed on soft foods (Ungar *et al.*, 2008). Comparative isotope analyses, in contrast, identified *P. boisei* as a grass-eater like the gelada baboon, *Theropithecus gelada* (Cerling *et al.*, 2011a). Overall dento-cranial morphology and the lack of shearing crests would have made it difficult for *P. boisei* to break down grasses however, save for some soft fresh shoots (Kay, 1975). Grasses are broken down by a scissor-like action of opposing molars. Teeth of *P. boisei* are low-crown and relatively flat when worn, which would have made it nearly impossible for them to break down tough grasses. An even stronger argument against grass-eating comes from broader biological considerations, specifically the species´ relatively large brain and its associated life history (Robson and Wood, 2008). Brains are expensive to grow and to maintain, and a diet consisting of mainly grasses is unlikely to fulfil the energetic requirements of a large-brained primate (Navarrete *et al.*, 2011). Unsurprisingly, the specialised grass-eating *Theropithecus gelada* is characterised by relatively and absolutely smaller brains compared to other baboons (Isler *et al.*, 2008). South African *P. robustus* is even more encephalised than *P. boisei* (Robson and Wood, 2008), which implies that its energy requirements may have even been higher than those of *P. boisei*.

Paranthropus robustus from South Africa is commonly considered to form a clade with East African *P. boisei* or, alternatively, to have evolved its derived morphology in response to similar ecological pressures, i.e., through homoplasy (Wood and Constantino, 2007). Either way, both *P. robustus* and *P. boisei* would therefore be expected to exhibit similar feeding habits. Apparently they do not. Isotope analyses imply that *P. robustus* consumed greater amounts of C_3 sources than *P. boisei*: it overlaps entirely in its isotope composition with *Australopithecus africanus* (Sponheimer *et al.*, 2005; Lee-Thorp *et al.*, 2010), but not in microwear texture (Scott *et al.*, 2005). Inspection of microwear textures provided by Grine *et al.* (2012) reveals *P. robustus* to be unique amongst extant and extinct primates, i.e., it does not unequivocally cluster with any other extant primate analysed thus far. This is intriguing, as is the observation that there is no overlap in microwear

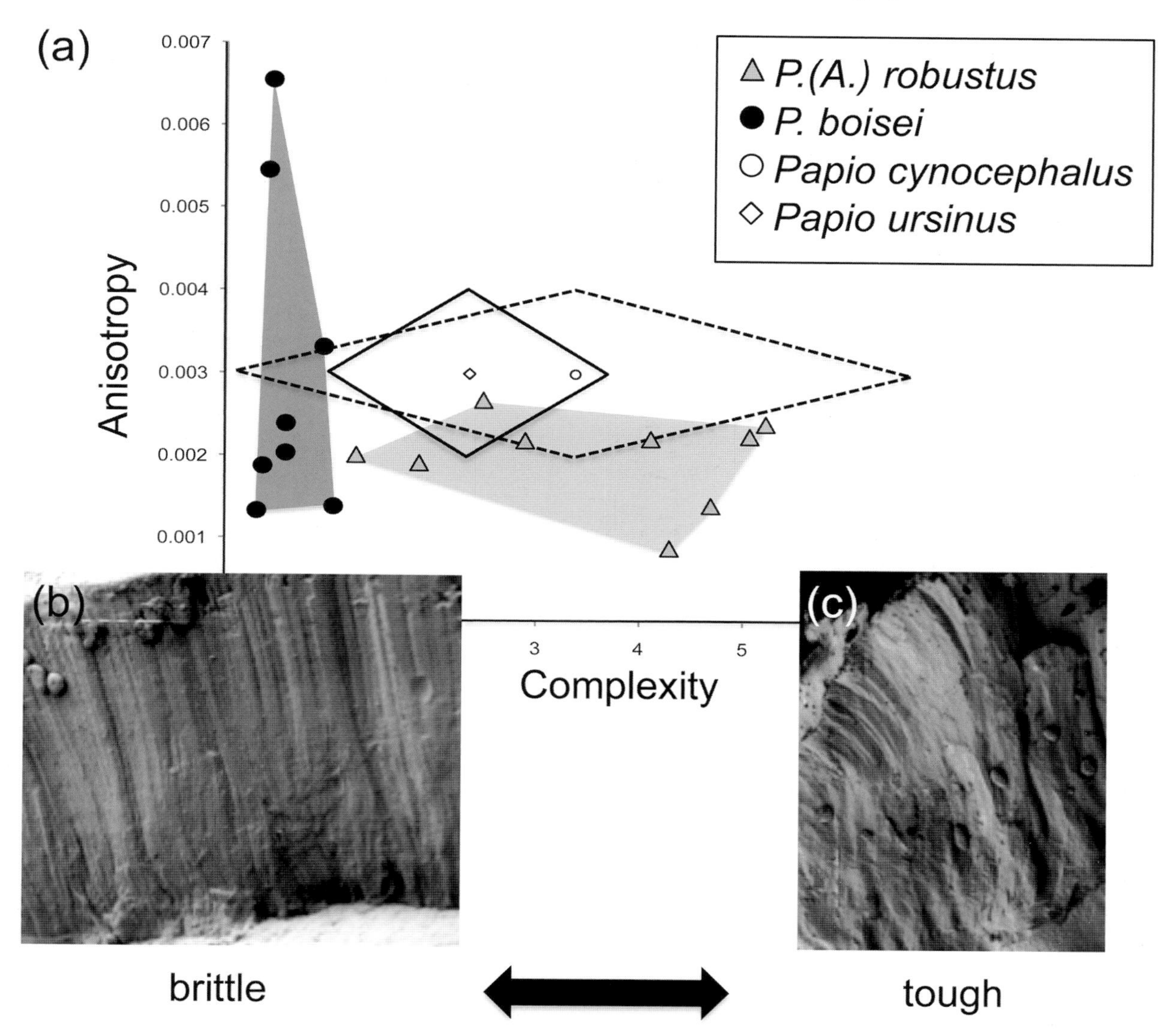

Figure 1.1. (a) The microwear texture data epLsar (anisotropy) and Asfe (complexity) were derived from Grine and colleagues (2012), whereby the data for papionins were originally presented in El-Zaatari et al. *(2005). (b) High resolution SEM picture of a naturally broken* P. boisei *tooth OH30. (c) SEM image of* P. robustus *SK55b. In* P. boisei *(b) enamel prisms are mostly straight and are organised nearly parallel. This renders the material stiff when loaded in the direction of the prisms, but weak when loaded perpendicular to it (see Figure 1.2). Consequently,* P. boisei *are poorly adapted to multi-directional loading. In contrast, prisms of* P. robustus *enamel undulate in a complex manner in 3D. Because of the different orientation of bundles of prisms, cracks would not propagate easily through the tissue. As a result,* P. robustus *teeth can be inferred to have been strong and well adapted to multi-directional loading. Information obtained from microwear textures, which do not overlap between the species (a), further highlight the distinctiveness in dietary ecology between paranthropines. Yet, each of these exinct hominin exhibits some similarities with baboons.*

textures between *P. robustus* and *P. boisei* at all (Figure 1.1a).

2.2. Biomechanics and Teeth

Enamel thickness continues to feature prominently in paleaoanthropological studies, not least because all hominins, particularly paranthropines, are characterised by hyperthick enamel. It is unclear however whether the thick enamel of these primates is an adaptation to wear resistance or to hard object feeding. To shed light on this question researchers commonly take a comparative approach whereby the correlation between diet in extant taxa and the enamel thickness of their teeth is used to make inferences about the dietary niche(s) of hominins (e.g. Kay, 1981; Dumont, 1995; Rabenold and Pearson, 2011). Such research designs are problematic, as extinct taxa do not (normally) have modern analogs. An assessment of the

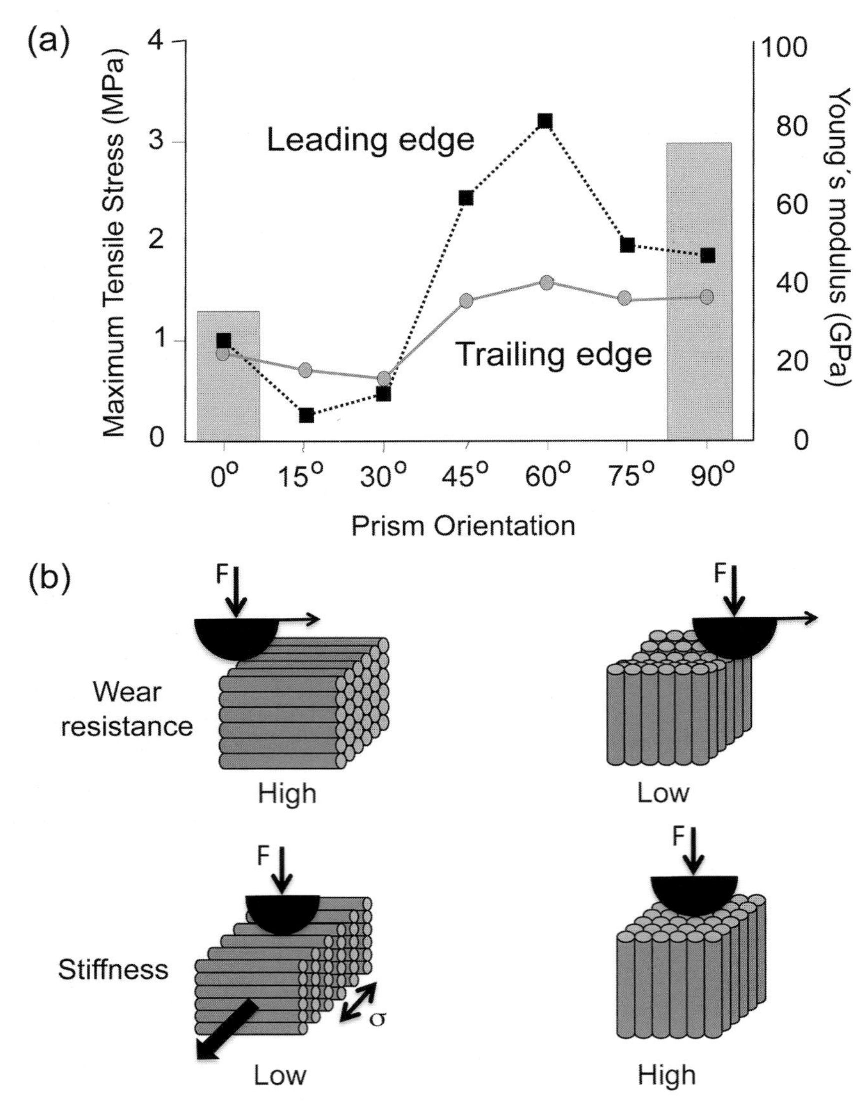

Figure 1.2. The accumulation of tensile stresses underneath the wear surfaces obtained in a finite element study, for the leading and trailing edge respectively (Shimizu et al., *2005). When the prisms are oriented parallel to the force vector (i.e., 0°–30°) wear resistance is high, but the dental material is soft and weak (i.e., tensile stresses building up between the prisms would push the tissue apart). At higher angles of prism orientation the tooth is stiff, but wear resistance is low. These opposite mechanical behaviours are schematically shown in (b). The schematic enamel blocks are positioned underneath the respective values shown in (a).*

micro- and macrostructural detail of enamel may be more informative, as the microanatomy determines whether the structure was adapted to resist abrasion, high bite forces and/or multi-directional loading.

Enamel is a composite hierarchically-organised material that is made up of a complex arrangement of hydroxyapatite crystals, both within and between prisms. Within prisms, i.e., the prism head, the crystal orientation is parallel with the prism long axis (*c*-axis), whereas in the prism tail/interprismatic matrix crystals are aligned nearly perpendicular to those of the prism head (Osborn, 1981); the change in crystal orientation from prism head to tail (interprismatic matrix) is gradual however. The diameter of a prism varies and is generally between 3–6μm (Figure 1.1b, c). Hydroxyapatite crystals are held together by an inorganic matrix and each prism is partly surrounded by a protein-rich prism sheath, which bonds the prisms and confers strength to the tissue (Ge *et al.*, 2005). The situation is further complicated by the fact that prisms are not (normally) parallel. Rather, prisms follow an undulating path from the dentino-enamel junction to the outer enamel surface, which varies in amplitude and frequency between taxa, and each prism is slightly offset with regard to its neighbour (Macho *et al.*, 2003). This results in bundles of prisms apparently crossing over, i.e., decussating (Figure 1.1c), when viewed in histological cross-sections (even though the transition is, in fact, gradual). Decussation acts as a crack-stopping mechanism (Rensberger, 1995, 2000). It is this complex hierarchical 3D arrangement of crystals within prisms, together with differences of prism orientations, that determines the biomechanical behaviour of the tissue (An *et al.*, 2012).

As crystals are considerably stiffer than the matrix, loads applied along the direction of the crystals will in large part be carried by the crystals: enamel will be stiff (Figure 1.2b). In contrast, when loads are applied across the direction of crystals, most of the internal stresses will be carried by the inorganic matrix and the prism sheaths. In this case, tensile stresses will build up between crystals/prisms and will push the tissue apart unless stopped by differently oriented crystals/prisms, i.e., prism decussation. Yet, it is in this latter situation that enamel is most resistant to wear (Rensberger and von Koenigswald, 1980; Maas, 1994; Shimizu *et al.*, 2005). In other words, wear resistance is least when the prisms approach the surface at a high angle, i.e., when the enamel is stiff, and greatest when the prisms approach the surface at a low angle and, hence, the tissue is prone to failure. Consequently, thick enamel *per se* can never be optimally designed to resist both abrasion *and* high loading. Prism organisation will (almost always) reflect a compromise between these two functions, although one function will dominate, depending on the properties of foods habitually consumed. Which aspect it is, i.e., adaptation to wear resistance or loading, can be deduced from inspection of enamel organisation with regard to overall prism decussation (Macho *et al.*, 2003, 2005; Macho and Shimizu, 2010) and prism orientation underneath the wear surface (Macho and Shimizu, 2009) (Figure 1.2).

Because mastication of tough foods requires a greater lateral excursion of the mandible, as well as a more varied angle of mandibular approach (Foster *et al.*, 2006; Woda *et al.*, 2006), the dental tissue needs to be protected against crack initiation and propagation resulting from multi-directional loading. The levels of prism undulation/decussation will therefore be high when the habitual diet consists of tough foods. This strengthening of the tooth can even be appreciated in the optical manifestations of prism decussation, the Hunter-Schreger bands (Kawai, 1955), but is best studied in naturally broken tooth surfaces using Scanning Electron Microscopy (Figure 1.1b, c). Conversely, mastication of hard, brittle and soft foods requires predominantly vertical forces, and no protective reinforcement against multidirectional loading is needed. In this case, little prism undulation/decussation is expected and prisms should ideally be facing cuspally for the force vector to be oriented parallel to the *c*-axis of prisms (Figure 1.1b). This information can be derived from enamel microstructure and, together with overall tooth morphology, allows inferences to be made about the stresses that habitually occur during mastication, as well as the levels of wear resistance.

Grasses are abrasive due to their high concentration of phytoliths. They are broken down in shear through a scissor-like action by opposing molars. Consumption of large quantities of grasses selects for thin enamel in order to retain shearing crests throughout life (Kay and Hiiemae, 1974). In order to counteract the rapid rate of wear, and to prolong the lifespan of the tooth, such teeth are high-crowned, i.e., hypsodont (Janis and Fortelius, 1988). This morphology is observed in the grass-eating *Theropithecus*, for example (Jablonski, 1993). As teeth of grass-eaters are also subjected to laterally directed loads due to the shearing action that is required to break down the tough grasses, enamel, although relatively thin, exhibits substantial amounts of re-inforcement in the form of prism decussation (Macho *et al.*, 1996) as well as a broad dental base and lateral buttress. This buttress ensures that masticatory loads are directed into the dental tissues when the tooth is loaded at an angle (Khera *et al.*, 1990; Macho and Spears, 1999). Gorillas, although frugivorous, also habitually feed on abrasive, tough leaves and herbacious matter. They therefore exhibit relatively high-cusped teeth with thin decussated enamel and a pronounced lateral buttress (Macho *et al.*, 2005; Macho and Shimizu, 2010). Lateral loading of straight-walled teeth would result in the force vector to cross the enamel cap, thereby causing shear along the loaded tooth wall and bending on the opposite side. This potentially causes transverse fractures (Rensberger, 2000). While *P. robustus* teeth have a lateral buttress, those of *P. boisei* do not (Macho, n.d.), the implications of which are discussed below.

2.3. Extant Baboons

In an innovative study, Altmann (1998) meticulously recorded the feeding ecology of yearling baboons, *Papio cynocephalus*, from the Amboseli National Park, Kenya. The C_4 foods eaten by these animals are varied and include corms, stolons, meristems, leaves, fruits, flowers, larvae, invertebrates, vertebrates and eggs. These items occur in different abundance throughout the year and, importantly, are of different nutritional value (Figure 1.3); for some foods the material properties are known also and allow judgments to be made as to whether the masticatory apparatus of hominins would have been capable of processing them (Dominy *et al.*, 2008). Because of their great dietary breadth compared with the great apes, baboons are better able to adjust their feeding behaviour to local conditions, to respond to seasonal fluctuations in resources and, consequently, to occupy a diverse range of habitats (Alberts *et al.*, 2005). Given the evolutionary success of our lineage, it is reasonable to assume that large-bodied large-brained hominins would have done the same. To test this proposition the feeding strategy of yearling baboons was analysed further (Altmann, 1998), but only C_4 sources were selected. First, the volume of foods eaten by the 2.27kg baboons (Altmann, 1998) was scaled up to be compatible with the larger sizes of hominins, ranging from 29–59kg: the volume of food consumed increases allometrically with body mass as $V_d=3.676M_b^{0.919}$ (Ross

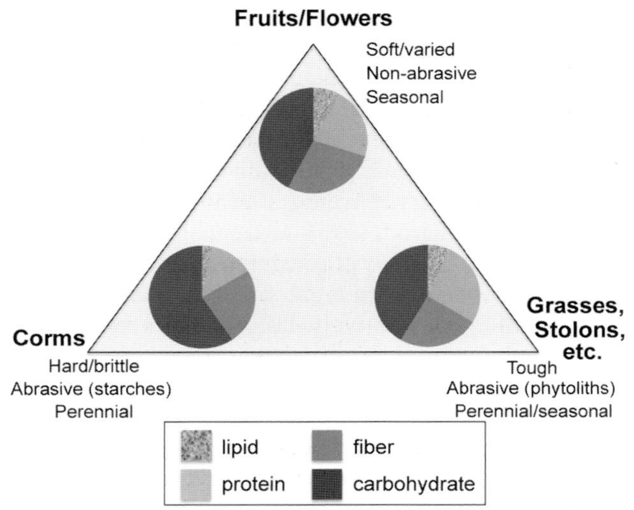

Figure 1.3. Information about C_4 plant foods eaten by Papio cynocephalus *were culled from Altman (1998), where the nutritional values can also be found. The foods are ordered according to the material properties whereby corms (i.e., hard and brittle) occupying the left-hand bottom corner, and hard abrasive foods are being placed in the right-hand corner. Soft items, like fruits and flowers, are clustered at the top. The average nutritional value for each category was calculated and is shown in the graph. In general, the nutrional yield of all foods selected by yearling baboons is high, whereby corms are highest in carbohydrates, and fresh grasses are highest in protein.*

et al., 2009). The relative proportion of different C_4 foods selected by yearling baboons (n=21), e.g. grasses versus invertebrates etc., was retained throughout the analyses, as it was assumed that baboons "intuitively know" which combination of foods maximizes nutritional return and minimises toxins (Altmann, 2009). Total feeding time on C_4 sources was then incrementally increased up to a maximum of 6 hours per day. Juvenile baboons only spend some 88 minutes per day feeding on C_4 foods and still derive a large amount of energy from their mother´s milk. In order to account for the isotopic composition of hominin tissue, which suggests a diet of 75%–80% of C_4 sources (Cerling *et al.*, 2011a), longer feeding bouts on C_4 resources must have taken place. What remains unclear is whether large-bodied hominins could have met their energetic requirements by feeding almost exclusively on C_4 foods, and whether hominins could have done so within the time budget available for feeding, i.e., maximum 50% of the day (6 hours).

Figure 1.4 shows the nutritional yield when the time for feeding on C_4 foods is increased, and for a range of body masses (because of uncertainties in body mass estimates of extinct hominins, no detailed predictions are attempted here). Owing to the high nutritional value of some C_4 foods selected by baboons, an increase in the volume consumed, i.e., through body mass and feeding time, leads to a sharp allometric increase in nutritional yield. The maximum energy output obtained for a 59kg hominin feeding on C_4

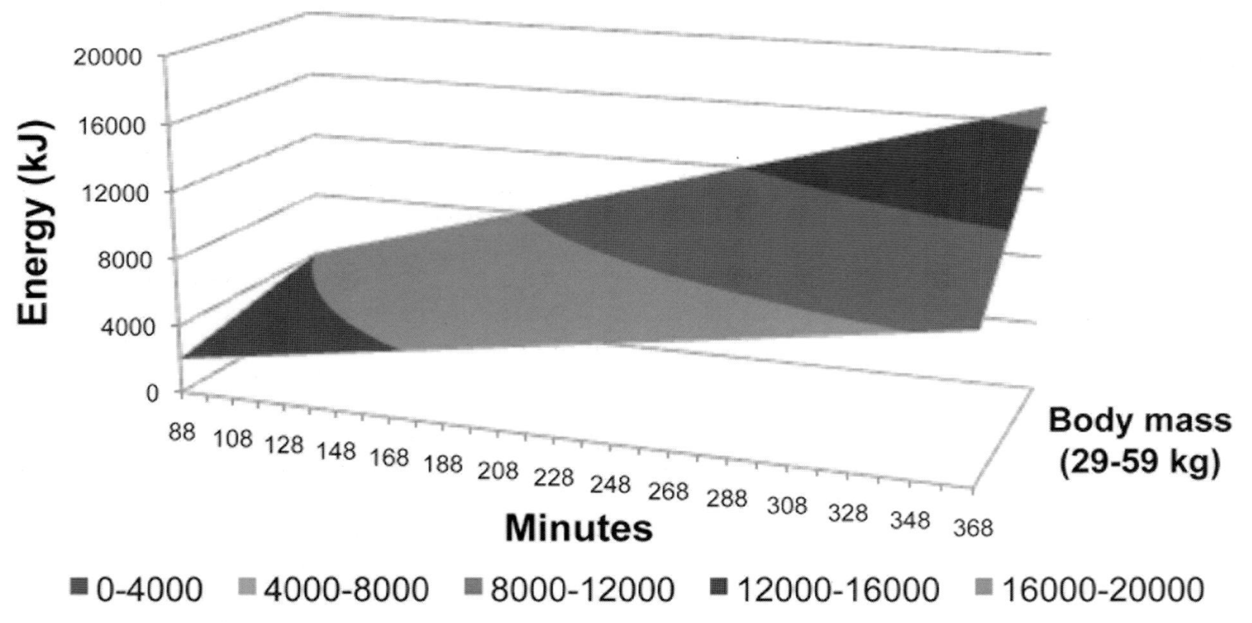

Figure 1.4. A simple energetic model was created from empirical data of yearling Papio cynocephalus *from the Amboseli National Park, Kenya. The nutritional yields (y-axis) are plotted for different primate body masses (z-axis) and feeding times. This serves to illustrate the overall trend, rather than to make specific predictions about a hominin at a given body mass. Also, this model does not take into account dento-cranial specialisations of hominins (e.g., enlarged tooth crown areas) or their greater manipulatory capabilities of adult baboons/hominins compared with juvenile baboons; incorporation of these factors would increase the nutritional returns even further.*

foods for 6 hours per day was over 16000 kJ (Figure 1.4). This is clearly within the time budget of a primate, as the daily energy expenditure of a primate is about 2–4 times its basal metabolic rate BMR (Pontzer and Kamilar, 2009). In reality, the values calculated here are most probably underestimations. The greater strengths of adult baboons/hominins for the extraction of below-ground resources and the improved manipulatory capabilities for cleaning and peeling USOs, when compared with juvenile baboons, have not been taken into account in the present calculations. Differences in tooth crown areas (Wood, 1991) and, hence, masticatory performance, have also not been considered. Yet, despite the conservative nature of the estimates, the outcomes clearly indicate that the selection of C_4 sources undertaken by *Papio cynocephalus*, could provide baboons (and hominins) with sufficient amounts of energy, protein and lipids, thus facilitating growth, development and maintenance of large brains (Navarette *et al.*, 2011). Importantly, the energetic output presented in Figure 1.4 demonstrates that the enriched $\delta^{13}C$ of hominin hard tissue need not be due to dietary specialisation, i.e., stenotopy, as the spectrum of C_4 foods eaten by baboons is varied and encompasses some 21 different types of foods (Figure 1.3). Nor is it justified to assume that hominins consumed a low-quality diet like other large-bodied grazing mammals (Cerling *et al.*, 2011a). What is more, a shift from a C_3 to a C_4 (predominantly) plant niche need not have incurred a change in total time budget for feeding. The C_4 plant niche is diverse and nutrient-rich (Figure 1.3), while the cognitive and manipulatory capabilities of hominins would have enabled them to fully explore and utilise its breadth. Subtle differences in behavioural strategies and ecological settings probably suffice to account for the morphological adaptations of various hominins, including those of *P. boisei* and *P. robustus*. Morphology may provide the clue as what these strategies were.

3. Dietary Ecology of *Paranthropus*

Hominins evolved in an increasingly open and fragmented environment (Cerling *et al.*, 2011b) and have long been presumed to have departed from their frugivorous ancestor to include a greater variety of foods, including grasses, sedges and USOs (Jolly, 1970; Sponheimer *et al.*, 2005). The latter vary greatly in material properties (Dominy *et al.*, 2008). Based on dento-cranial morphology it is reasonable to infer that the dietary strategies of all hominins differed, even though their isotope composition may have been the same (Cerling *et al.*, 2013). *Paranthropus boisei* and *P. robustus* are a case in point.

All South African hominins overlap in $\delta^{13}C$ values with papionins (Codron *et al.*, 2008) despite differences in dento-cranial morphology between hominins and papionins, on the one hand, and between *A. africanus* and *P. robustus*, on the other. This further strengthens arguments for similarities in dietary strategies and radiations among baboons and hominins (Jolly, 2001). The thick enamel of *P. robustus* teeth and their cuspally aligned enamel prisms imply consumption of hard and abrasive foods (Macho and Shimizu, 2009). Unlike *P. boisei* however, the enamel of *P. robustus* is highly decussated (Figure 1.1c), while teeth exhibit lateral buttresses. This morphology is consistent with consumption of tough foods, like stolons, tubers etc., but not grasses: like *P. boisei*, teeth of *P. robustus* lack shearing crests.

Paranthropus boisei teeth, in contrast, are steep-walled and lack substantial amounts of prism decussation (Figure 1.1b); the lack of decussation has previously been surmised on the basis of Hunter-Schreger bands (Beynon and Wood, 1986) but can be more fully appreciated in SEM images (Figure 1.1b, c). Overall, teeth of *P. boisei* are therefore ill-equipped to masticate tough foods, i.e., dissipate laterally directed loads. Instead, hard brittle foods, like corms, and invertebrates and soft fruits could be easily broken down. Fruits are available only seasonally and would therefore not have been part of the animal's staple, while the quantities of invertebrates that can be eaten is also limited. Instead, owing to their high nutritional value and sweet taste, corms constitute core foods for baboons (Altmann, 1998; Pochron, 2000; Swedell *et al.*, 2007), are liked by modern humans and are sometimes used as staple foods (e.g. Belewu and Belewu, 2007; Bamishaiye and Bamishaiye, 2011). Importantly, they can be eaten raw. In baboons, an increased intake of corms was found to lead to high rates of tooth wear (Galbany *et al.*, 2011). The thick enamel and blunt cusps, i.e., bunodont, teeth of *P. boisei* are optimally designed to slow down the process of rapid wear. However, the high rate of wear is not (predominantly) a consequence of phytoliths contained within these corms, but starches.

Starches are highly abrasive in unheated state (Singh *et al.*, 2003). Microwear textures resulting from masticating corms are expected to be indistinct, i.e., uninformative, though. This is because starches are not broken down mechanically, but chemically through the interaction with salivary amylase (Perry *et al.*, 2007). Lengthy oral processing, i.e., milling, facilitates this chemical breakdown, whereby the tooth surfaces become "polished" by starches being randomly moved across the tooth surfaces. As a consequence, microwear textures will mimick those found in soft fruit eaters (Ungar *et al.*, 2008). It is parsimonious to suggest that the high starch content of corms (Umerie *et al.*, 1997; Umerie and Ezeuzo 2000) probably explains the high rates of tooth wear seen in *P. boisei*, the flat pattern of its mesowear (Tobias, 1967) and the indistinctive microwear textures (Ungar *et al.*, 2008). Cranial morphology is compatible with such an interpretation too, although with some re-interpretation. Corms are hard, but not excessively so. However, milling of these starch-rich foods requires a repetitive chewing motion, which would place demands on the masticatory musculature, if performed regularly. With this in mind, suggestions that the masticatory apparatus of *P. boisei* is an adaptation to repetitive chewing, rather than to habitual employment of high bite forces, seem supported (Demes and Creel, 1988). Modifications observed in the temporomandibular joint that facilitate the transverse movement of the mandible (du Brul, 1977) are in line with such propositions too, as are palaeoecological considerations. Many sedges and grasses, particularly of the family *Cyperus* and *Poaceae,* produce corms and thrive in, but are not exclusive to, marshlands and edaphic grasslands commonly associated with *P. boisei* (Bonnefille, 2010). Although palynological data for South African sites are largely wanting, it is evident that these hominins occupied a different vegetation zone (Bonnefille, 2010). This must have affected the dietary niche available to *P. robustus*. It is therefore unclear whether ecological differences alone could explain the different dietary strategies of East and South African australopiths or, alternatively, whether the morphologies reflect true differences in dietary preferences among hominins. Regardless, based on dental anatomies, it is parsimonious to conclude that South African *P. robustus* exploited food sources that were hard, tough and abrasive, e.g. stolons, tubers etc., whereas *P. boisei* consumed large quantities of hard, brittle and abrasive foods, like corms. Some overlap in food choices can nonetheless be expected between these two closely-related hominins, given that they were probably opportunistic omnivores like baboons.

In conclusion, it is likely that the C_4 foods selected by hominins derived from the same broad spectrum of C_4 plants that is exploited by *Papio* today. The fact that hominins fed on these nutrient-rich sources for longer periods of time and, consequently, consumed larger quantities thereof (Cerling *et al.*, 2013) suffices to account for the isotope composition of their hard tissue. This shift in preference may have been crucial for the evolutionary pathway of hominins, including their marked changes in brain size and life histories. Rather than being of low quality, many C_4 sources, particularly those chosen by extant baboons, are of high nutritional value. The fact that these foods are ubiquitous and that competition for these resources with other mammals is limited, probably contributed further to the evolutionary success of our lineage.

Acknowledgements

I thank the Trustees of the National Museums of Tanzania and Kenya, The Transvaal Museum and the University of the Witwatersrand for access to fossil material. Funding was provided by the Ministerio de Ciencia e Innovación (CGL2010-20868). Thanks to Amanda Korstjens for providing detailed feedback on the ideas expressed in this contribution.

References

Alberts, S.C., Hollister-Smith, J.A., Mututua, R.S., Sayialel, S.N., Muruthi, P.M., Warutere, J.K., Altmann J. 2005. Seasonality and long-term change in a savannah environment, in: Brockmann, D.K., van Schaik, C.P. (Eds.), *Seasonality in Primates*. Cambridge University Press, Cambridge, pp. 157–195.

Altmann, S.A. 1998. *Foraging for Survival.* University of Chicago Press, Chicago, IL.

Altmann, S.A. 2009. Fallback foods, eclectic omnivores, and the packaging problem. *American Journal of Physical Anthropology* 140, 615–629.

An, B., Wan, R., Zhang D. 2012. Role of crystal arrangement on the mechanical performance of enamel. *Acta Biomaterialia* 8, 3784–3793.

Bamishaiye, E.I., Bamishaiye, O.M. 2011. Tiger nut: as a plant, its derivatives and benefits. *African Journal of Food, Agriculture, Nutrition and Development* 11, 5157–5170.

Belewu, M.A., Belewu, K.Y. 2007. Comparative physico-chemical evaluation of tiger-nut, soybean and coconut milk sources. *International Journal of Agriculture and Biology* 9, 785–787.

Beynon A.D., Wood, B.A. 1986. Variations in enamel thickness and structure in East African hominids. *American Journal of Physical Anthropology* 70, 177–193.

Bonnefille, R. 2010. Cenozoic vegetation, climate change and hominid evolution in tropical Africa. *Global and Planetary Change* 72, 390–411.

du Brul, E.L. 1977. Early hominid feeding mechanism. *American Journal of Physical Anthropology* 47, 305–320.

Carmody, R.N., Weintraub, G.S., Wrangham, R.W. 2011. Energetic consequences of thermal and nonthermal food processing. *Proceedings of the National Academy of Sciences USA* 108, 19199–19203.

Cerling, T.E., Manthi, F.K., Mbua, E.N., Leakey, L.N., Leakey, M.G., Leakey, R.E., Brown, F.H., Grine, F.E., Hart, J.A., Kaleme, P., Roche, H., Uno, K.T., Wood, B.A. 2013. Stable isotope-based diet reconstructions of Turkana Basin hominins. *Proceedings of the National Academy of Sciences USA* 110, 10501–10506.

Cerling, T.E., Mbua, E., Kirera, F.M., Manthi, F.K., Grine, F.E., Leakey, M.G., Sponheimer, M., Uno, K.T. 2011a. Diet of *Paranthropus boisei* in the early Pleistocene of East Africa. *Proceedings of the National Academy of Sciences USA* 108, 9337–9341.

Cerling, T.E., Wynn, J.G., Andanje, S.A., Bird, M.I., Korir, D.K., Levin, N.E., Mace, W., Macharia, A.N., Quade, J., Remien, C.H. 2011b. Woody cover and hominin environments in the past 6 million years. *Nature* 476, 51–56.

Codron, D., Lee-Thorp, J.A., Sponheimer, M., de Ruiter, D., Codron, J. 2008. What insights can baboon feeding ecology provide for early hominin niche differentiation? *International Journal of Primatology* 29, 757–772.

Dumont, E.R. 1995. Enamel thickness and dietary adaptation among extant primates and chiropterans. *Journal of Mammalogy* 76, 1127–1136.

Demes, B., Creel, N. 1988. Bite force, diet, and cranial morphology of fossil hominids. *Journal of Human Evolution* 17, 657–670.

Dominy, N.J., Vogel, E.R., Yeakel, J.D., Constantino, P., Lucas, P.W. 2008. Mechanical properties of plant underground storage organs and implications for dietary models of early hominins. *Journal of Evolutionary Biology* 35, 159–175.

Elton, S. 2006. Forty years on and still going strong: the use of hominin-cercopithecid comparisons in palaeoanthropology. *Journal of the Royal Anthropological Institute* 12, 19–38.

El-Zaatari, S., Grine, F.E., Teaford, M.F., Smith, H.F. 2005. Molar microwear and dietary reconstructions of fossil

Ceropithecoidea from the Plio-Pleistocene deposits of South Africa. *Journal of Human Evolution* 49, 180–205.

Foster, K.D., Woda, A., Peyron, M.A. 2006. Effect of texture of plastic and elastic model foods on the parameters of mastication. *Journal of Neurophysiology* 95, 3469–3479.

Galbany, J., Altmann, J., Pérez-Pérez, A., Alberts, S.C. 2011. Age and individual foraging behavior predict tooth wear in Amboseli baboons. *American Journal of Physical Anthropology* 144, 51–59.

Ge, J., Cui, F.Z., Wang, X.M., Feng, H.L. 2005. Property variations in the prism and the organic sheath within enamel by nanoindentation. *Biomaterials* 26, 3333–3339.

Grine, F.E., Sponheimer, M., Ungar, P.S., Lee-Thorp, J., Teaford, M.F., 2012. Dental microwear and stable isotopes inform the paleoecology of extinct hominins. *American Journal of Physical Anthropology* 148, 285–317.

Isler, K., Kirk, E.C., Miller, J.M.A., Albrecht, G.A., Gelvin, B.R., Martin, R.D. 2008. Endocranial volumes of primate species: scaling analyses using a comprehensive and reliable data set. *Journal of Human Evolution* 55, 967–978.

Hatley, T., Kappelman, J. 1980. Bears, pigs, and Plio-Pleistocene hominids: A case for the exploitation of belowground food resources. *Journal of Human Ecology* 8, 371–387.

Jablonski, N.G. (Ed.) 1993. *Theropithecus. The Rise and Fall of a Genus*. Cambridge, Cambridge University Press.

Janis, C.M., Fortelius, M. 1988. On the means whereby mammals achieve increased functional durability of their dentitions, with special reference to limiting factors. *Biological Revues of the Cambridge Philosophical Society* 63, 197–230.

Jolly, C.J. 1970. The seed eaters: A new model of hominid differentiation based on a baboon analogy. *Man* 5, 5–26.

Jolly, C.J. 2001. A proper study for mankind: Analogies from the papionin monkeys and their implications for human evolution. *Yearbook of Physical Anthropology* 44, 177–204.

Kay, R.F. 1975. The functional adaptations of primate molar teeth. *American Journal of Physical Anthropology* 43, 195–215.

Kay, R.F. 1981. The nut-crackers: A new theory of the adaptations of the Ramapithecidae. *American Journal of Physical Anthropology* 55, 141–151.

Kay, R.F., Hiiemae, K.M. 1974. Jaw movement and tooth use in recent and fossil primates. *American Journal of Physical Anthropology* 40, 227–256.

Kawai, N. 1955. Comparative anatomy of the bands of Schreger. *Okajimas Folia Anatomica Japonica* 27, 115–131.

Khera, S.C., Carpenter, C.W., Vetter, J.D., Staley, R.N. 1990. Anatomy of cusps of posterior teeth and their fracture potential. *Journal of Prosthetic Dentistry* 64, 139–147.

Lee-Thorp, J., 2011. The demise of "Nutcracker Man". *Proceedings of the National Academy of Sciences USA* 108, 9319–9320.

Lee-Thorp, J.A., Sponheimer, M., Passey, B.H., de Ruiter, D.J., Cerling, T.E. 2010. Stable isotopes in fossil hominin tooth enamel suggest a fundamental dietary shift in the Pliocene. *Philosophical Transactions of the Royal Society* 365, 3389–3396.

Leonard, W.R., Robertson, M.L. 1997. Comparative primate energetics and hominid evolution. *American Journal of Physical Anthropology* 102, 265–281.

Maas, M.C. 1994. A scanning electron-microscopic study of *in vitro* abrasion of mammalian tooth enamel under compressive loads. *Archives of Oral Biology* 39, 1–11.

Macho G.A. (n.d.) The implications of morphology, mechanics and microstructure of teeth for understanding dietary drivers in human evolution, in: Lee-Thorp, J.A., Katzenberg, M.A. (Eds.), *Handbook of the Archaeology of Diet*. Oxford University Press, Oxford, in press.

Macho, G.A., Jiang, Y., Spears, I.R. 2003. Enamel microstructure – a truly three-dimensional structure. *Journal of Human Evolution* 45, 821–830.

Macho, G.A., Reid, D.J., Leakey M.G., Jablonski N., Beynon, D.A. 1996. Climatic effects on dental development of *Theropithecus oswaldi* from Koobi Fora and Olorgesailie. *Journal of Human Evolution* 30, 57–70.

Macho, G.A., Shimizu, D. 2009. Dietary niches of South African australopiths: Inference from enamel prism attitude. *Journal of Human Evolution* 57, 241–247.

Macho, G.A., Shimizu, D. 2010. Kinematic parameters inferred from enamel microstructure: New insights into the diet of *Australopithecus anamensis*. *Journal of Human Evolution* 58, 23–32.

Macho, G.A., Shimizu, D., Jiang, Y., Spears, I.R. 2005. *Australopithecus anamensis*: A finite element approach to studying functional adaptations in extinct hominins. *The Anatomical Record* 283A, 310–318.

Macho, G.A., Spears, I.R. 1999. The effects of loading on the biomechanical behaviour of molars of *Homo, Pan* and *Pongo*. *American Journal of Physical Anthropology* 109, 211–227.

Navarrete, A., van Schaik, C.P., Isler, K. 2011. Energetics and the evolution of human brain size. *Nature* 480, 91–94.

O'Connell, J.F., Hawkes K., Blurton Jones, N.G. 1999. Grandmothering and the evolution of *Homo erectus*. *Journal of Human Evolution* 36, 461–485.

O'Connell, J.F., Hawkes, K., Lupo, K.D., Blurton Jones, N.G. 2002. Male strategies and Plio-Pleistocene archaeology. *Journal of Human Evolution* 43, 831–872.

Osborn, J.W. (Ed.) 1981. *Dental Anatomy and Embryology*. Blackwell Scientific Publications, Oxford.

Perry, G.H., Dominy, N.J., Claw, K.G., Lee, A.S., Fiegler, H., Redon, R., Werner, J., Villanea, F.A., Mountain, J.L., Misra, R., Carter, N.P., Lee, C., Stone, A.C. 2007. Diet and the evolution of human amylase gene copy number variation. *Nature Genetics* 39, 1256–1260.

Pochron, S.T. 2000. The core dry-season diet of Yellow baboons (*Papio hamadryas cynocephalus*) in Ruaha National Park, Tanzania. *Folia Primatologica* 71, 346–349.

Pontzer, H, Kamilar, J.M. 2009. Great ranging associated with greater reproductive investment in mammals. *Proceedings of the National Academy of Sciences USA* 106:192–196.

Rabenold, D., Pearson, O.M. 2011. Abrasive, silica phytoliths and the evolution of thick molar enamel in primates, with implications for the diet of *Paranthropus boisei*. *PlosONE* 6, e28379.

Rensberger, J.M. 1995. Determination of stresses in mammalian enamel and their relevance to the interpretation of feeding behaviors in extinct taxa, in: Thomason, J.J. (Ed.), *Functional Morphology in Vertebrate Paleontology*, Cambridge University Press, Cambridge, pp. 151–172.

Rensberger, J.M. 2000. Pathways to functional differentiation in mammalian enamel, in: Teaford, M.F., Smith, M.M., Ferguson, M.W.J. (Eds.), *Development, Function and Evolution of Teeth*. Cambridge University Press, Cambridge, pp. 252–268.

Rensberger, J.M., von Koenigswald, W. 1980. Functional and phylogenetic interpretation of enamel microstructure in rhinoceroses. *Paleobiology* 6, 477–495.

Robson, S.L., Wood, B. 2008. Hominin life history: Reconstruction and evolution. *Journal of Anatomy* 212, 394–425.

Ross, C.F., Washington, R.L., Eckhardt, A., Reed, D.A., Vogel, E.R., Dominy, N.J., Machanda, Z.P. 2009. Ecological consequences of scaling of chew cycle duration and daily feeding time in Primates. *Journal of Human Evolution* 56, 570–585.

Scott, R.S., Ungar, P.S., Bergstrom, T.S., Brown, C.A., Grine, F.E., Teaford, M.F., Walker, A. 2005. Dental microwear texture analysis shows within-species diet variability in fossil hominins. *Nature* 436, 693–695.

Shimizu, D., Macho, G.A., Spears, I.R. 2005. The effect of prism orientation and loading direction on contact stresses in prismatic enamel: Implications for interpreting wear patterns. *American Journal of Physical Anthropology* 126, 427–434.

Singh, N., Singh, J, Kaur, L, Sodhi, N.S., Gill, B.S. 2003. Morphological, thermal and rheological properties of starches from different botanical sources. *Food Chemistry* 81, 219–231.

Sponheimer, M., Lee-Thorp, J., de Ruiter, D., Codron, D., Codron, J., Baugh, A.T., Thackeray, F. 2005. Hominins, sedges, and termites: new carbon isotope data from the Sterkfontein valley and Kruger National Park. *Journal of Human Evolution* 48, 301–312.

Swedell, L., Hailemeskel, G., Schreier, A. 2007. Composition and seasonality of diet in wild Hamadryas baboons: Preliminary findings from Filoha. *Folia Primatologica* 79, 476–490.

Tobias, P.V. 1967. *The Cranium and Maxillary Dentition of Australopithecus (Zinjanthropus) boisei* (Olduvai Gorge). Cambridge University Press, Cambridge.

Umerie, S.C., Ezeuzo, H.O. 2000. Physicochemical characterization and utilization of *Cyperus rotundus* starch. *Bioresource Technology* 72, 193–196.

Umerie, S.C., Obi, N.A.N., Okafor, E.O. 1997. Isolation and characterization of starch from *Cyperus esculentus* tubers. *Bioresource Technology,* 62, 63–65.

Ungar, P.S., Grine, F.E., Teaford, M.F. 2008. Dental microwear and diet of the Plio-Pleistocene hominin *Paranthropus boisei. PlosONE* 3, e2044.

Woda, A., Foster, K., Mishellany, A., Peyron, M.A. 2006. Adaptation of healthy mastication to factors pertaining to the individual or to the food. *Physiology & Behavior* 89, 28–35.

Wood, B. 1991. *Koobi Fora Research Project*. Vol. 4. Clarendon Press, Oxford.

Wood, B., Constantino, P. 2007. *Paranthropus boisei:* Fifty years of evidence and analysis. *Yearbook of Physical Anthropology* 50, 106–132.

2. Recording Primate Spinal Degenerative Joint Disease Using a Standardised Approach

Diana Mahoney Swales and Pia Nystrom

Clinical and palaeopathological studies of non-human primates have identified lesions morphologically identical to those ascribed to degenerative joint disease in humans. However, no standardised criteria for recording such changes in non-human primates have been established, limiting direct comparisons of studies between human and non-human primate skeletal material. This research analysed the applicability of current recording criteria for spinal degenerative joint disease (SDJD) in archaeological populations to studies of extant non-human primate species. The distribution and type of lesions exhibited by arboreal, semi-arboreal, and terrestrial primates were compared directly, allowing the impact of locomotor activity upon SDJD to be investigated.

*Macroscopic degenerative changes to the vertebrae of several baboon species (*Papio spp.*), blue monkeys (*Cercopithecus mitis*) and greater spot-nosed monkeys (*Cercopithecus nictitans*) were recorded using criteria common in human palaeopathological studies. The results demonstrate that such criteria can be applied to primate skeletal remains with positive results. A much higher prevalence and severity of SDJD in terrestrial relative to arboreal and semi-arboreal primates indicates a probable link between substrate, locomotion and patterns of degeneration of the spine. An earlier onset and higher prevalence of SDJD in the baboon sample may also suggest a possible association with body mass.*

Keywords: Primate; Spondyloarthropathy; Palaeopathology; Comparative analysis

1. Introduction

Spinal degenerative joint disease (SDJD) is a common condition observed in the human skeletal archaeological record (Aufderheide and Rodríguez-Martín, 1998; Weiss and Jurmain, 2007). Indeed, it appears to have been an affliction for the entirety of hominid existence, with australopithecines (Cook *et al.*, 1983), Neanderthals (Trinkaus, 2005) and early modern humans (Morris *et al.*, 1987) exhibiting vertebral degenerative lesions. Previous clinical and palaeopathological studies of non-human primates, particularly apes and old world monkeys, have reported the presence of degenerative joint disease and osteoarthritic lesions morphologically identical to those seen in modern humans (DeRousseau, 1985; Rothschild and Woods, 1992; Carlson *et al.*, 1994; Jurmain, 2000; Nakai, 2003). This indicates that SDJD is an affliction of all primate species, not just those that are habitually bipedal. However, it has been recognized in studies of both human (Bridges, 1993) and non-human (Lovell, 1991) primates that inconsistencies in the way SDJD is recorded has been detrimental to studies of this condition.

It is an inevitability of advancing age that the joints gradually deteriorate (Roberts and Manchester, 1995). Several studies have shown that degeneration of the vertebral column in humans is age progressive, typically commencing in the fourth decade of life and intensifies in prevalence and severity thereafter (Jurmain and Kilgore, 1995; Aufderheide and Rodríguez-Martín, 1998; Ortner 2003). SDJD appears also to be age progressive in both captive (DeRousseau, 1985) and wild (Lovell, 1990, 1991) non-human primate species. However, the aetiology of human degenerative joint disease, particularly osteoarthritis, has been attributed to a number of additional contributing factors to advancing age, such as sex, body mass, bone mineral content, trauma, genetic predisposition, biomechanical stress, physical activity

(Burr *et al.*, 1983; Jurmain and Kilgore, 1995; Rogers and Waldron, 1995; Brickley and Waldron, 1998; Jurmain, 2000; Kahl and Smith, 2000; Šlaus, 2000; Knüsel, 2003; Ortner, 2003; Weiss and Jurmain, 2007; Rojas-Sepulveda *et al.*, 2008), and the evolution of bipedality (Jankauskas, 1992; Knüsel *et al.*, 1997; Weber *et al.*, 2003).

The current study attempts to evaluate the applicability of standardised recording criteria for SDJD in humans to non-human primates. The standardised criteria will be applied to arboreal (blue monkeys and spot nosed monkeys) and terrestrial (baboons) primates to investigate the impact of locomotor activity, substrate, body size, sex and age upon the expression of SDJD in a sample population. This project was essentially a pilot study to identify the validity of such a methodology and identify future avenues of research.

Many studies of human SDJD follow a reasonably simple diagnostic criterion of slight, moderate and severe modelled upon the guidelines provided by Stewart (1958, reproduced in Aufderheide and Rodríguez-Martín, 1998), Buikstra and Ubelaker (1994) and Steckel *et al.* (2006). Variation in how the data are quantified and presented results in conflicting reports on the relationship of SDJD between different primate species and between non-human and human primates. For example, some studies report a higher frequency of SDJD lesions in humans relative to other primates (Jurmain, 1989, 2000), whereas in others the reverse is documented (Rothschild and Woods, 1991; Kramer *et al.*, 2002) and in some, humans and primates exhibit similar levels of SDJD (Nuckley *et al.*, 2008).

The research presented here assesses the applicability of the commonly utilised standards for recording osteophytosis, porosity and eburnation in human skeletal remains (Buikstra and Ubelaker, 1994; Steckel *et al.*, 2006) to primate skeletal material. Furthermore, newly developed methodologies for quantifying the severity of Schmorl's nodes (Üstündağ, 2009) and ossification of the supraspinous, longitudinal and yellow ligaments (Hukuda *et al.*, 2000) are applied to a primate assemblage. The distribution and type of lesions exhibited by arboreal, semi-arboreal, and terrestrial primates are compared directly using crude and true frequencies of lesions (definitions of prevalence rates are provided below), allowing the impact of locomotor activity upon the distribution and manifestation of degenerative joint disease in the vertebral column to be investigated.

The objective of this research is to transcend the boundaries of human osteology, biological and evolutionary anthropology and primatology to better understand the complex aetiology of SDJD in extant and extinct humans. Implementing a standardised method of data collection and a comparison of different primate species enables the creation of a reference line for patterns of DJD in wild and captive primate species, facilitating a greater comprehension of the influence of factors such as social group and behaviour, body size, age, locomotor patterns and physical environment and ecology (Gillespie *et al.*, 2008) on the evolution and aetiology of SDJD in fossil and living human and non-human primates and osteoarchaeological assemblages.

2. Background Information

Several macroscopically identifiable skeletal modifications were employed to identify and classify SDJD, namely osteophyte formation at the joint margins, and porosity and eburnation of the joint surface. Osteophytes, ossifications of the fibrocartilage extending horizontally from the anterior and lateral margins of the vertebral body (*spondylosis deformans*) develop in response to narrowing of the intervertebral space and subsequent periosteal reactions as the disc degenerates (Aufderheide and Rodríguez-Martín, 1998). Further bony outgrowths (syndesmophytes) develop, predominantly on the lateral margins to support joints that have become unstable or weakened by age and physical activity related stress (Rogers *et al.*, 1987; Rogers and Waldron, 1995; Knüsel *et al.*, 1997; Aufderheide and Rodríguez-Martín, 1998; Van der Merwe *et al.*, 2006). Porosity (osteochondrosis) on the articular surface is often linked to degenerative joint disease, particularly osteoarthritis, as it is associated with degeneration of the hyaline cartilage and is thus utilised as a diagnostic criterion (Buikstra and Ubelaker, 1994; Knüsel *et al.*, 1997; Steckel *et al.*, 2006). Eburnation on the joint surface results simply from the friction once the joint cartilage has been compromised and insufficient cartilage survives to prevent contact between the opposing joint surfaces (Rothschild, 1997). The presence of eburnation on the apophyseal joints is often cited as being pathognomic of osteoarthritis, typically in the literature associated with appendicular joints (Rogers and Waldron, 1995), however, it has also been argued to simply represent a more advanced stage of joint degeneration (Rothschild, 1997; Molnar *et al.*, 2011). Either way, the presence of eburnation enables identification of the specific degenerative condition of vertebral osteoarthritis or the severity of SDJD (Buikstra and Ubelaker, 1994; Waldron, 2009). A further identifying characteristic of spinal degeneration and the deleterious effect of ageing on the vertebral column is ossification of the supraspinous, longitudinal and yellow ligaments (Hukuda *et al.*, 2000). Even though the aetiology of ossification of the longitudinal and yellow ligments is unknown, an association has been made between these ligamentous ossifications and the syndesmophyte formation observed in diffuse idiopathic skeletal hyperostosis (Resnick *et al.*, 1978; Hukuda *et al.*, 2000). It is probable that the ligaments ossify in response to the instability and weakening of the joint with advancing age (as stated earlier for osteophyte formation). This, and positive correlations between the development of musculo-skeletal markers and the presence of eburnation in the appendicular joints in human archaeological assemblages (Molnar *et al.*, 2011) deems the inclusion of the ossification of the spinal ligaments beneficial to enhancing the understanding of the aetiology and development of the SDJD process.

Schmorl's nodes, circular depressions on the cranial and/or caudal surfaces of the vertebral body created by the erosive pressures resulting from herniation of the *nucleus pulposa* when the intervertebral disc prolapses (Faccia and Williams, 2008), are common in extant and archaeological human populations, but are rarely reported in other animals and non-human primates (Lovell, 1991; Fews *et al.*, 2006). Schmorl's nodes have been ascribed several aetiologies, including congenital predisposition and trauma, but have also been attributed to age related degeneration due to their close association with other forms of degenerative joint disease (Buikstra and Ubelaker, 1994; Üstündağ, 2009). Schmorl's nodes have in the past been stated as being an age related phenomenon frequently afflicting individuals aged over 45 years (Aufderheide and Rodríguez-Martín, 1998), which strongly suggests they result from age related degeneration of the intervertebral disc. However, more recent palaeopathological and clinical literature has identified no definite correlation between the frequency and severity of Schmorl's nodes with advancing age, but often identify a link between degeneration of the intervertebral disc and the vertebral bone adjacent to the lesion (Šlaus, 2000; Pfirrmann and Resnick, 2001; Plomp *et al.*, 2012). Their possible association with degenerative joint disease and high presence in human spinal palaeopathology warrants the inclusion of Schmorl's nodes in the current study.

Osteoarthritis, the most commonly documented degenerative joint disease in archaeological skeletal assemblages, has often been used in the literature as an indicator of comparative physiological stress within and between populations and of the type of subsistence economy and lifestyle they experienced (Jurmain, 1990; Croft *et al.*, 1992; Jurmain and Kilgore, 1995; Rojas-Sepulveda *et al.*, 2008). Indeed, the primary explanation for variation in the prevalence of osteoarthritis and degenerative joint disease is biomechanical stress imposed upon the joints related to occupational and subsistence strategies (Rogers and Waldron, 1995; Klaus *et al.*, 2009; Eshed *et al.*, 2010). However, within archaeological reports on human assemblages, there is often no distinction made between osteoarthritis and non-specific SDJD (Bridges, 1993; Lovell, 1994), and the diagnostic criteria for identifying osteoarthritis are not always described. Consequently, it is not possible to confirm the presence of eburnation, the pathognomic diagnostic feature of osteoarthritis (Rogers and Waldron, 1995), or to determine the prevalence of osteoarthritis and SDJD as separate conditions. The prevalence rates of typical degenerative changes of the symphyseal and appositional surfaces of the vertebrae were combined and documented within this study under the term spinal degenerative joint disease, but the fact that the lesions grouped under this heading could have several aetiologies was kept in mind.

3. Materials and Method

A number of methods have been employed to record human and non-human primate SDJD with varying success (Lovell, 1991; Bridges, 1993; Knüsel *et al.*, 1997; Kramer *et al.*, 2002), but since the publication of Buikstra and Ubelaker's *Standards for Data Collection from Human Skeletal Remains* (1994), and the shift from case study reviews to more comparative inter- and intra-population studies (Larsen, 2002; Steckel and Rose, 2002; Roberts and Cox, 2003; Steckel *et al.*, 2006) there has been an increasing requirement for standardised recording criteria for human skeletal assemblages. Comparisons of human collections using standardised recording methods and prevalence rates have provided informative results about past human populations (Knüsel *et al.*, 1997; Roberts and Cox, 2003; Klaus *et al.*, 2009). Therefore, it is hypothesised that the application of such an approach to non-human primates would be equally informative.

Macroscopic osteological analysis was undertaken on the vertebral columns of three primate species selected for variability in their habitual locomotion and postural behaviour, and typical living environment (Table 2.1). The skeletons of 12 baboons, comprising olive (*Papio anubis*), yellow (*P. cynocephalus*), hamadryas (*P. hamadryas*) and chacma (*P. ursinus*) baboons represent the terrestrial quadrupedal sample. The skeletal remains of 10 blue monkeys (*Cercopithecus mitis stuhlmanni*) constitute the arboreal sample and the semi-arboreal primates were represented by four spot-nosed monkey (*Cercopithecus nictitans)* skeletons. All the primates were wild specimens curated at the Natural History Museum, London and the Duckworth Collection at the Powell Cotton Museum, Dorset.

Baboons are terrestrial quadrupeds, which inhabit semi desert, savannah, scrubland and wood environments. Males and females exhibit notable sexual dimorphism with the male weight ranging from 16–32kg and female weight ranging between 9–18kg (Jaffe and Isbell, 2011). Blue monkeys are arboreal quadrupeds, which inhabit evergreen, semi-deciduous woodland and bamboo and dry scrub. The males (5.9–8.9kg) are generally larger than the females (3.83kg) (Jaffe and Isbell, 2011). Spot-nosed monkeys are semi-arboreal quadrupeds, which live in the rainforest. The males are larger in size than the females (males 6.6kg, females 3.65–4.2kg) (Jaffe and Isbell, 2011).

Each primate analysed was a subadult (close to reaching skeletal maturity) or adult of known sex (Table 2.1). The stage of epiphyseal union of the post-cranial skeleton was recorded to determine between adult and immature individuals. Each primate was assigned an age of subadult, young, prime and old adult, based on dental development and occlusal surface wear (Phillips-Conroy *et al.*, 1988, 2000). It is difficult to assign an exact age at death for primates. However, the use of wide, but progressive, age categories for each primate species derived from occlusal wear does provide an age approximation sufficient in this

Table 2.1. Summary data of primates analysed and the expression of SDJD lesions.

Species	Common Name	I.D.	Origin	Captive	Sex	Age	Osteophytes	Porosity	Eburnation	Schmorl's Nodes
C. mitis stuhlmanni	Blue monkey	ZD.1972.68	Uganda	Wild	M	YA	0	0	0	0
C. mitis stuhlmanni	Blue monkey	ZD.1972.72	Kenya	Wild	M	PA	0	0	0	0
C. mitis stuhlmanni	Blue monkey	ZD.1972.83	Kenya	Wild	F	OA	0	0	0	0
C. mitis stuhlmanni	Blue monkey	ZD.1972.88	Kenya	Wild	M	YA	0	0	0	0
C. mitis stuhlmanni	Blue monkey	ZD.1972.76	Kenya	Wild	M	A	0	0	0	0
C. mitis stuhlmanni	Blue monkey	ZD.1972.82	Kenya	Wild	M	OA	1	1	1	0
C. mitis stuhlmanni	Blue monkey	ZD.1972.87	Kenya	Wild	F	OA	1	1	1	0
C. nictitans	Greater spot-nosed	M-310	Cameroon	Wild	M	YA	0	0	0	0
C. nictitans	Greater spot-nosed	M-792**	Cameroon	Wild	F	SA	0	0	0	0
C. nictitans	Greater spot-nosed	M-336	Cameroon	Wild	M	OA	0	0	0	0
C. nictitans	Greater spot-nosed	M-433	Cameroon	Wild	M	OA	0	0	0	0
P. anubis	Olive baboon	ZD.1901.8.9.23	Kenya	Wild	F	OA	1	1	0	0
P. anubis	Olive baboon	ZD.1962.12.14.6	Kenya	Wild	F	YA	1	0	0	0
P. anubis	Olive baboon	ZD.1935.2.14.1	Uganda	Wild	M	PA	1	1	1	0
P. anubis	Olive baboon	ZD.1973.1291	Tanzania	Wild	M	OA	1	0	0	0
P. anubis	Olive baboon	ZD.1862.6.26.1	Angola	Wild	M	A	1	1	0	0
P. cynocephalus	Yellow baboon	ZD.1972.129	Kenya	Wild	M	PA	1	0	0	0
P. cynocephalus	Yellow baboon	ZD.1962.7.6.13	Kenya	Wild	M	PA	1	0	0	0
P. cynocephalus	Yellow baboon	ZD.1972.130	Unknown	Wild	F	YA	1	1	0	0
P. cynocephalus	Yellow baboon	ZD.1948.3.30.1	Unknown	Wild	F	YA	1	0	0	0
P. hamadryas	Hamadryas	ZD.1980.394	Ethiopia	Wild	M	PA/SA	1	1	0	0
P. ursinus	Chacma baboon	ZD.1857.12.21.2*	unknown	Wild	M	SA	1	1	1	0
P. ursinus	Chacma baboon	ZD.1973.12.90	S. Africa	Wild	M	PA	1	0	0	0

instance to identify the relationship between the stage in the lifecycle and degeneration of the spine. It is unlikely that the levels of dental occlusion are notably affected by extreme dietary differences or non-masticatory activities within the primate species. Therefore, the assignment of adolescent, young adult, prime adult, and old adult is consistent with human studies, due to the inaccuracy of osteological ageing methodology.

The degenerative changes under observation included osteophyte formation, eburnation, porosity of the joint surface, and ligamentous ossification, as described previously. Recording of SDJD was undertaken predominantly using the criteria devised by Steckel *et al.* (2006), with additional observations of Schmorl's nodes and ligamentous ossification recorded following the requirements of Üstündağ (2009) and Hukuda *et al.* (2000), respectively. Trauma was recorded if visible upon a vertebra or if it was present upon another skeletal element and was contributory to the presence of SDJD. This enabled a distinction to be made between idiopathic degeneration and that resulting as a secondary response to trauma, infectious, metabolic, congenital, neurological or circulatory disease (Aufderheide and Rodríguez-Martín, 1998). Any cases of possible secondary degenerative joint disease resulting from a traumatic episode or infection were excluded from the study.

Degenerative changes in the spine were recorded for each vertebra. Each component, i.e., the cranial and caudal discal surfaces of the vertebral body, the articular processes, and costal articular facets were analysed for signs of degeneration. The degree of osteophytosis, porosity and eburnation were scored using a numerical scale (0 = element not present; 1 = element present, but no degenerative changes; 2 = slight; 3 = moderate; 4 = severe). Scores 2 to 4 correspond with Groups I to III illustrated by Brothwell (1981) and Grades 1 to 3 described by Steckel *et al.* (2006). In this instance the criteria designed by Buikstra and Ubelaker (1994) were not employed because it was difficult to consistently implement descriptions such as 'elevated rings', 'pinpoint' and 'coalesced' and it was decided that the type of lesion and percentage of joint surface affected were associated, and therefore would not be separated into two separate categories.

Schmorl's nodes were recorded using the descriptions and photographs provided by Üstündağ (2009), which are adapted from Knüsel *et al.* (1997). Any Schmorl's node covering less than half of the cranial or caudal surface of the vertebral body, with a depth less than 2mm was recorded as 'slight', and any lesions deeper and/or larger than that were classified as 'severe'. The lesions were accorded a similar numeric scale as employed for recording osteophytosis, porosity and eburnation (0 = element not present; 1 = element present, but no lesions present; 2 = slight; 3 = severe).

The distributions of the lesions, which may be important for determining the aetiology of modifications such as osteophyte formation, were recorded on diagrammatic representations of each vertebra. Such a simplified recording system, using clearly defined criteria based on pictorial and photographic representations of joint disease, was employed to ensure a consistent level of recording and to enable replication of the methods in future studies.

Ossification of the anterior longitudinal ligament, posterior longitudinal ligament and yellow ligament were scored as 'probable' or 'definite' according to the criteria proposed by Hukuda *et al.* (2000) and drawn upon the relevant vertebra on the recording form.

Increasingly, in the human palaeopathological literature comparative analysis between different archaeological populations has been enabled by the use of crude and true prevalence rates, whereby the frequency of disease is calculated as a percentage of the total number of individuals within a population presenting one or more elements that could possibly be effected (CPR), and a percentage of the actual number of elements, i.e., number of joint surfaces, teeth or bone diaphysis (TPR), respectively (Waldron, 1994, pp. 45, 54). Within this study, the crude prevalence rate (number of affected vertebrae divided by the number of individuals with recordable vertebrae) was calculated for each lesion type (osteophyte, porosity, eburnation, Schmorl's nodes, and ligament ossification) for each species. The true prevalence was also calculated for each individual and each vertebral category, namely the number of recordable elements (cervical, thoracic and lumbar vertebrae) exhibiting lesions characteristic of degenerative joint disease divided by the total number of recordable elements.

High crude prevalence rates (CPR) for SDJD are consistently observed for archaeological populations, with sites from the Neolithic through to the late medieval period revealing a crude prevalence of approximately 50.0% or more (Roberts and Cox, 2003). Prevalence rates have been deemed a reliable approximator for non-fatal conditions, such as SDJD (Waldron and Rogers, 1991), but the use of average CPR for an archaeological population is likely to underestimate the occurrence of any disease as lesions may not be recordable on all individuals, a problem compounded by the friable nature of vertebral bodies in archaeological deposits. Consequently, such high percentages of cases are a clear indicator of the ubiquity of SDJD in the past.

4. Results

The primates examined during this study exhibited marginal osteophyte formation, porosity and eburnation lesions identical to those observed in modern humans (Tables 2.2 and 2.3). However, none of the primates examined exhibited Schmorl's nodes, or ossification of the spinal ligaments.

The arboreal blue monkey and semi-arboreal spot-nosed monkey exhibited limited degeneration of the spinal column. None of the spot-nosed monkeys showed any evidence for spinal degenerative joint disease. Only barely discernible marginal osteophytes were seen on the

Table 2.2. Frequencies and percentages of spinal degenerative joint disease lesions in the primate assemblage.

Species	n	Osteophytes		Porosity		Eburnation		Schmorl's Nodes	
		n	%	n	%	n	%	n	%
Cercopithecus mitis	7	2	28.57	2	28.57	2	28.57	0	0.00
Cercopithecus nictitans	4	0	0.00	0	0.00	0	0.00	0	0.00
Papio spp	1	12	100.00	6	50.00	2	16.67	0	0.00

Table 2.3. Frequencies and percentages of SDJD lesions in each age category in the primate assemblage as a percentage of individuals in each age category.

Species	Age	n	Osteophytes		Porosity		Eburnation	
			n	%	n	%	n	%
Cercopithecus mitis	young adult	2	0	0.00	0	0.00	0	0.00
	prime adult	1	0	0.00	0	0.00	0	0.00
	old adult	3	2	66.66	2	66.66	2	66.66
Papio spp	subadult	2	2	100.00	1	50.00	1	50.00
	young adult	2	2	100.00	1	50.00	0	0.00
	prime adult	5	5	100.00	2	40.00	1	20.00
	old adult	2	2	100.00	1	50.00	0	0.00

Table 2.4. Frequencies and percentages of SDJD lesions in males and females in the primate assemblage.

Species	Sex	n	Osteophytes		Porosity		Eburnation	
			n	%	n	%	n	%
Cercopithecus mitis	Male	5	1	20.00	1	20.00	1	20.00
	Femal	2	1	50.00	1	50.00	1	50.00
Papio spp	Male	8	8	100.00	4	50.00	2	25.00
	Femal	4	4	100.00	2	50.00	0	0.00

vertebral bodies of the two blue monkeys. In contrast, all of the terrestrial baboons exhibited osteophyte formation on the margins of the vertebral bodies, of which half also exhibited porosity of either the body or the costal facets. A further two of the twelve baboons exhibited eburnation of the superior articular facets and transverse processes. The presence of eburnation on the apophyseal joints confirms the presence of osteoarthritis as it is recognised in the palaeopathological and osteoarchaeological literature.

The onset of SDJD occured later in blue monkeys than baboons (Table 2.3). The only two blue monkeys exhibiting SDJD were older adult individuals, whereas the affected baboons included all age categories. Indeed, the greatest frequency of eburnation in the baboon sample was observed amongst the sub-adults and prime adults.

Amongst the blue monkeys, females exhibited a greater frequency of osteophytes, porosity and eburnation than males (Table 2.4; females 1/2, males 1/5). This is most probably because the two blue monkey females are both older adults, whereas the male assemblage ranges from young adults through to one older adult (Table 2.1). Within the baboon sample there are equal percentages of both males and females exhibiting osteophytosis and porosity. The two affected individuals are a prime and subadult males.

Amongst the baboon sample the cervical, thoracic and lumbar vertebrae are all affected by osteophytic lipping, whereas only one upper cervical, lower thoracic and lumbar vertebrae are affected in the blue monkey sample (Figure 2.1). The higher true prevalence (TPR) in the baboon sample further emphasises the greater affliction of this species with SDJD relative to the blue monkeys.

5. Discussion

The authors are conscious that the sample size is too small to enable viable statistical analysis and that the trends and their interpretation presented herein are tentative. However, the purpose of this paper is to identify the viability of the current methods for recording human SDJD to primate assemblages to enable further research.

Locomotor behaviour and substrate use appears to be the prime contributing factor in the prevalence of SDJD in the non-human primates included in this study. The low prevalence of spinal degeneration observed in the arboreal and semi-arboreal primates, relative to the terrestrial baboons and human populations, is consistent with the primate literature (Bramblett, 1967; Alexander, 1989, 1994; Rothschild and Woods, 1992; Jurmain, 2000). The substrate

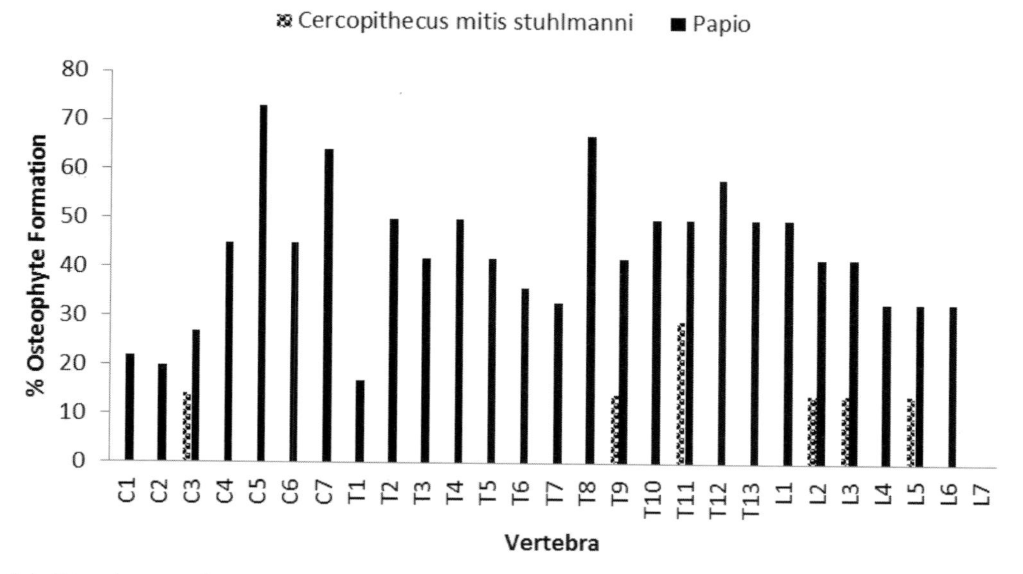

Figure 2.1. Distribution of marginal osteophytes throughout the vertebral columns of blue monkeys and baboons.

within and upon which the primates are locomoting appears to impact on levels of SDJD. The higher prevalence in terrestrial baboons may reflect the inflexible nature of the hard ground they encounter imposing greater stress upon the skeleton, whereas arboreal substrates absorb a greater proportion of the impact of locomotion.

SDJD does appear to be age progressive in the non-human primates examined in this study. The greater prevalence of SDJD amongst the older adult female blue monkeys relative to the males, which were represented by all age categories, indicates age related degeneration to be a principal cause of such lesions in this species. The presence of eburnation in the elderly blue monkeys only, suggests it to be an age progressive indicator of SDJD and represents a more severe stage of degeneration. There is no distinction between the frequencies of SDJD recorded for the male and female baboons. Both sexes are represented by all of the age groups, i.e. there is no age bias, as seen in the blue monkeys. Consequently, it appears that SDJD, manifest as marginal osteophytes and porosity on the joint surface, is age progressive in both arboreal and terrestrial non-human primates and not sex specific.

In baboons there appears to be a male bias for eburnation lesions. The occurrence of eburnation in the prime and subadult male baboons and contrasting absence amongst the females indicate a male predilection for this condition. Both baboon skeletons exhibited eburnation alongside systemic degenerative lesions in the form of osteophytes on most of the surviving appendicular joint surfaces. This may reflect advancing age or recurrent high impacts on all the joint surfaces, further suggesting that polishing on the joint surface and the associated loss of the joint cartilage represent an advanced stage of degenerative joint disease, and that SDJD and osteoarthritis can be regarded as the same pathological condition. It is possible that the eburnation expressed in the prime and subadult male baboons could result from increased physical stress or trauma to the cartilage during terrestrial locomotion resulting from their larger body size. The proposed influence of body size upon the greater expression of SDJD is supported by two observations. Firstly, both males and females in the larger baboon species, which exhibits the greater and most frequent expression of SDJD lesions, have a greater body weight than either the males or females in the two arboreal species. The baboons (including males and females) weigh 9–32kg on average, whereas the greatest weight for the male arboreal primates is the male blue nosed monkeys, which have an average maximum weight of 8.9kg. Secondly, the spot-nosed guenons, which exhibit no signs of SDJD, are the smallest of the three species. The positive relationship between body mass and expression of SDJD is consistent with the documented higher frequency of SDJD, particularly osteophyte formation, observed in mountain gorillas, relative to other great apes (Lovell, 1991).

The low prevalence of SDJD in the arboreal blue monkeys and spot nosed monkeys also corresponds with the hypothesis that arboreal primates employ a greater range of joint movements and therefore are not afflicted by SDJD resulting from under-use of the joints range of movements, i.e., they are not affected by the discrepancy in current function and inherited design of the joint (Alexander, 1989, 1994).

The differences in the distribution of degeneration throughout the vertebral column may indicate different stresses imposed throughout the spine within each species, or morphological differences.

The absence of Schmorl's nodes in the primate sample suggests their formation may be dependent upon stresses imposed on the cranial and caudal surfaces of the vertebral bodies during bipedal locomotion. The absence of Schmorl's nodes may reflect lower impact trauma to the intervertebral discs or a lack of reduction of the intervertebral space associated with age related degeneration in humans. Alternatively, their absence may reflect a physiological characteristic of the vertebrae or the cartilage, which prevents their formation.

Humans consistently show a set distribution of degenerative lesions throughout the vertebral column, which is commonly ascribed to the curvature of the human spine, which has co-evolved with bipedalism (Jurmain and Kilgore, 1995; Jurmain, 2000). Jurmain (2000) argues that the consistent pattern of distribution of SDJD in humans makes intra- and inter-population comparisons futile for evaluating behavioural, morphological and physiological differences amongst humans, but that comparisons with other primate species could provide a greater understanding of the effect of bipedality on the skeleton. A great consensus appears to be that aside from age, biomechanical stress is a prime cause of degenerative joint disease. The evidence presented here appears to support such a hypothesis and indicates that a reduction of the forces imposed upon the skeleton due to flexible substrate and reduced body size and mass reduces the formation of SDJD.

The use of a diagrammatic recording form, upon which the score for each proposed diagnostic characteristic (spondylosis deformans, osteochondrosis, eburnation and Schmorl's nodes) can be clearly identified for each skeletal component of the vertebra (body, superior and inferior processes, transverse processes and costal facets), enables comparisons to be made with observations in previously recorded material, which may have focused on particular lesions and manifestations of degenerative joint disease or used specific terminology. For example, the distinction made between vertebral and apophyseal osteophytosis (Hukuda *et al.*, 2000).

6. Conclusions

The main conclusion from this study is that human SDJD recording criteria can be applied to non-human primate skeletal material with positive results. Comparisons of the prevalence of SDJD between arboreal, semi-arboreal and terrestrial quadrupeds revealed a higher prevalence of spinal degeneration amongst the terrestrial primates, implicating substrate and locomotor behaviour as being influential upon the forces impacting the joint surface. A possible association between SDJD and increasing body mass was also identified. The eburnation observed in subadult and prime adult male baboons may result from increased trauma to the cartilage resulting from repetitive high impact activities, such as jumping or running for long distances, compounded by their larger body mass.

By testing the applicability of human recording criteria to primate samples this small study indicates the potential of a palaeoepidemiological study of spinal and appendicular degenerative joint disease in a larger range of primate species. Such a study could provide evidence which could help determine the evolution and epidemiology of SDJD.

Overall it is clear that methods and knowledge can be transferred between numerous fields and a positive shift towards collaborations and multi-disciplinary approaches between osteology, biological anthropology, primatology and palaeoanthropology alongside clinical research from the fields of human and animal medicine can enhance not only the information available to us, but also our abilities to interpret such information. The intention of this project and the proposed future developments is to update the current knowledge regarding degenerative joint disease and life history stressors in human and non-human primates and fossil hominids to enhance our understanding of the influence of environment, life cycle and lifestyle upon the skeleton. Specifically, it is the intention of the authors to continue examining the influence of substrate and positional behaviour using a wide range of primate species.

Acknowledgements

The authors would like to thank the Natural History Museum (London) and Powell Cotton Museum (Dorset) for allowing access to their collections. This study was made possible by a BABAO Academic Small Research Grant.

References

Alexander, C.J. 1989. Relationship between the utilisation profile of individual joints and their susceptibility to primary osteoarthritis. *Skeletal Radiology* 18, 199–205.

Alexander, C.J. 1994. Utilisation of joint movement range in arboreal primates compared with human subjects: An evolutionary frame for primary osteoarthritis. *Annals of Rheumatic Diseases* 53, 720–725.

Aufderheide, A.C., Rodríguez-Martín, C. 1998. *The Cambridge Encyclopedia of Human Paleopathology*. Cambridge University Press, Cambridge.

Bramblett, C.A. 1967. Pathology in the Darajani baboon. *American Journal of Physical Anthropology* 26, 331–340.

Brickley, M., Waldron, T. 1998. Relationship between bone density and osteoarthritis in a skeletal population from London. *Bone* 22, 279–283.

Bridges, P.S. 1993. The effect of variation in methodology on the outcome of osteoarthritic studies. *International Journal of Osteoarchaeology* 3, 289–295.

Brothwell, D.R. 1981. *Digging Up Bones*, 3rd edition. Cornell University Press, New York.

Buikstra, J.E., Ubelaker, D.H. 1994. S*tandards for Data Collection from Human Skeletal Remains. Proceedings of a Seminar at the Field Museum of Natural History.* Arkansas Archeological Survey Research Series No. 44, Fayetteville.

Burr, D.B., Bruce Martin, R., Schaffler, M.B., Jurmain, R.D., Harner, E.J., Radin, E.L. 1983. Osteoarthrosis: Sex-specific relationship to osteoporosis. *American Journal of Physical Anthropology* 61, 299–303.

Carlson, C.S., Loeser, R.F., Jayo, M.J., Weaver, D.S., Adams, M.R., Jerome, C.P. 1994. Osteoarthritis in cynomolgus macaques: A primate model of naturally occurring disease. *Journal of Orthopedic Research* 12, 331–339.

Cook, D.C., Buikstra, J.E., DeRousseau, C. J., Johanson, D.C. 1983. Vertebral pathology in the Afar australopithecines. *American Journal of Physical Anthropology* 60, 83–101.

Croft, P., Coggon, M., Cruddas, M., Cooper, C. 1992. Osteoarthritis of the hip: An occupational disease in farmers. *British Medical Journal* 304, 1269–1272.

DeRousseau, C.J. 1985. Aging in the musculoskeletal system of rhesus monkeys. II. Degenerative joint disease. *American Journal of Physical Anthropology* 67, 177–184.

Eshed, V., Gopher, A., Pinhasi, R., Hershkovitz, I. 2010. Paleopathology and the origin of agriculture in the Levant. *American Journal of Physical Anthropology* 143, 121–133.

Faccia, K.J., Williams, R.C. 2008. Schmorl's nodes: Clinical significance and implications for the bioarchaeological record. *International Journal of Osteoarchaeology* 18, 28–44.

Fews, D., Brown, P.J., Alterio, G.L.D. 2006. A case of invertebral disc degeneration and prolapse with Schmorl's node formation in a sheep. *Veterinary and Comparative Orthopaedics and Traumatology* 3, 187–189.

Gillespie, T.R., Nunn, C.L., Leendertz, F.H. 2008. Integrative approaches to the study of primate infectious disease: Implications for biodiversity conservation and global health. *American Journal of Physical Anthropology* 137, 53–69.

Hukuda. S., Inoue, K., Ushiyama, T., Saruhashi, Y., Iwasaki, A., Huang, J., Mayeda, A., Nakai, M., Li, F.X., Yang, Z.Q. 2000. Spinal degenerative lesions and spinal ligamentous ossifications in ancient Chinese populations of the Yellow River civilization. *International Journal of Osteoarchaeology* 10, 108–124.

Jaffe, K.E., Isbell, L.A. 2011. The guenons: Socioecology of polyspecific associations, in: Campbell, C.J., Fuentes, A., MacKinnon, K.C., Bearder, S.K., Stumpf, R.M. (Eds.), *Primates in Perspective*, 2nd edition. Oxford University Press, New York, pp. 277–300.

Jankauskas, R. 1992. Degenerative changes of the vertebral column in Lithuanian paleosteological material. *Anthropologie* XXX, 109–119.

Jurmain, R.D. 1989. Trauma, degenerative disease, and other pathologies among the Gombe chimpanzees. *American Journal of Physical Anthropology* 80, 229–237.

Jurmain, R.D. 1990. Paleoepidemiology of a central California prehistoric population from CA-Ala 329. II. Degenerative disease. *American Journal of Physical Anthropology* 83, 83–94.

Jurmain, R.D. 2000. Degenerative joint disease in African great apes: an evolutionary perspective. *Journal of Human Evolution* 39, 185–203.

Jurmain, R.D., Kilgore, L. 1995. Skeletal evidence of osteoarthritis: A palaeopathological perspective. *Annals of Rheumatic Diseases* 54, 443–450.

Kahl, K.E., Smith, M.O. 2000. The pattern of spondylosis deformans in prehistoric samples from west-central New Mexico. *International Journal of Osteoarchaeology* 10, 432–446.

Klaus, H.D., Larsen, C.P., Tam, M.E. 2009. Economic intensification and degenerative joint disease: life and labor on the postcontact north coast of Peru. *American Journal of Physical Anthropology* 139, 204–221.

Knüsel, C.J. 2003. On the biomechanical and osteoarthritic differences between hunter-gatherers and agriculturalists. *American Journal of Physical Anthropology* 91, 523–527.

Knüsel, C.J., Göggel, S., Lucy, D. 1997. Comparative degenerative joint disease of the vertebral column in the medieval monastic cemetery of the Gilbertine priory of St. Andrew, Fishergate, York, England. *American Journal of Physical Anthropology* 103, 481–495.

Kramer, P.A., Newell-Morris, L.L., Siinkin, P.A. 2002. Spinal degenerative disk disease (DDD) in female macaque monkeys: epidemiology and comparison with women. *Journal of Orthopaedic Research* 20, 399–408.

Larsen, C.S. 2002 Bioarchaeology: The lives and lifestyles of past people. *Journal of Archaeological Research* 10, 119–66.

Lovell, N.C. 1990. Skeletal and dental pathology of free-ranging mountain gorillas. *American Journal of Physical Anthropology* 81, 399–412.

Lovell, N.C. 1991. An evolutionary framework for assessing illness and injury in nonhuman primates. *Yearbook of Physical Anthropology* 34, 117–155.

Lovell, N.C. 1994. Spinal arthritis and physical stress at Bronze Age Harappa. *American Journal of Physical Anthropology* 93, 149–164.

Molnar, P., Ahlstrom, T.P., Leden, I. 2011. Osteoarthritis and activity – an analysisnof the relationship between eburnation, Musculoskeletal Stress Markers (MSM) and age in two Neolithic hunter-gatherer populations from Gotland, Sweden. *International Journal of Osteoarchaeology* 21, 283–291.

Morris, A.G., Thackeray, A.I., Thackeray, J.F. 1987. Late Holocene human skeletal remains from Snuifklip, near Vleesbaai, Southern Cape. *South African Archaeological Bulletin* 42, 153–160.

Nakai, M. 2003. Bone and joint disorders in wild Japanese macaques from Nagano Prefecture, Japan. *International Journal of Primatology* 24, 179–195.

Nuckley, D.J., Kramer, P.A., Del Rosario, A., Fabro, N., Baran, S., Ching, R.P. 2008. Intervertebral disc degeneration in a naturally occurring primate model: radiographic and biomechanical evidence. *Journal of Orthopaedic Research* 26, 1283–1288.

Ortner, D.J. 2003. *Identification of Pathological Conditions in Human Skeletal Remains*, 2nd edition. Academic Press, New York.

Pfirrmann, C.W.A., Resnick, D. 2001. Schmorl's nodes of the thoracic and lumbar spine: radiographic-pathologic study of prevalence, characterization, and correlation with degenerative changes of 1,650 spinal levels in 100 cadavers. *Radiology* 219, 368–174.

Phillips-Conroy, J.E., Jolly, C.J. 1988. Dental eruption schedules of wild and captive baboons. *American Journal of Primatology* 15, 17–29.

Phillips-Conroy, J.E., Bergman, T., Jolly, C.J. 2000. Quantitative assessment of occlusal wear and age estimation in Ethiopian and Tanzanian baboons, in: Whitehead, P.F., Jolly, C.J. (Eds.), *Old World Monkeys*. Cambridge University Press, Cambridge, pp. 321–340.

Plomp, K.A., Roberts, C.A., Strand Viðarsdóttir, U. 2012. Vertebral morphology influences the development of Schmorl's nodes in the lower thoracic vertebrae. *American Journal of Physical Anthropology* 149, 572–582.

Resnick, D., Guerra, J., Robinson, C.A., Vint, V.C. 1978. Association of diffuse idiopathic skeletal hyperostosis (DISH) and calcification and ossification of the posterior longitudinal ligament. *American Journal of Roentgenology* 131, 1049–1053.

Roberts, C., Cox, M. 2003. *Health and Disease in Britain*. Sutton Publishing, Stroud.

Roberts, C.A., Manchester, K. 1995. *The Archaeology of Disease*, 2nd edition. The History Press, Stroud.

Rogers, J., Waldron, T. 1995. *A Field Guide to Joint Disease in Archaeology*. John Wiley Publishing, New York.

Rogers, J., Waldron, T., Dieppe, P., Watt, I. 1987. Arthropathies in paleopathology. The basis of classification according to most probably cause. *Journal of Archaeological Science* 14, 179–193.

Rojas-Sepúlveda, C., Ardagna, Y., Dutour, O. 2008. Paleoepidemiology of vertebral degenerative disease in a Pre-Columbian Muisca series from Colombia. *American Journal of Physical Anthropology* 135, 416–430.

Rothschild, B.M. 1997. Porosity: A curiosity without diagnostic significance. *American Journal of Physical Anthropology* 104, 529–533.

Rothschild, B.M., Woods, R.J. 1991. Erosive arthritis in representative defleshed bones. *American Journal of Physical Anthropology* 85, 125–134.

Rothschild, B.M., Woods, R.J. 1992. Spondyloarthropathy as an Old World phenomenon. *Seminars in Arthritis and Rheumatism* 21, 306–316.

Šlaus, M. 2000. Biocultural analysis of sex differences in mortality profiles and stress levels in the late medieval population from Nova Rača, Croatia. *American Journal of Physical Anthropology* 111, 193–209.

Steckel, R.H., Larsen, C.S., Sciulli, P.W., Walker, P.L. 2006. *The Global History of Health Project: Data Collection Codebook*. Available from: http://global.sbs.ohio-state.edu/european_module.htm

Steckel, R.H., Rose, J.C. (Eds.) 2002. *The Backbone of History: Health and Nutrition in the Western Hemisphere*. Cambridge University Press, Cambridge.

Stewart, T.D. 1958. The rate of development of vertebral osteoarthritis in American whites and its significance in skeletal age identification. *The Leech* 28, 144–151.

Trinkaus, E., 2005. Pathology and the posture of the La Chapelle-aux-Saints Neandertal. *American Journal of Physical Anthropology* 67, 19–41.

Üstündağ, H. 2009. Schmorl's nodes in a post-medieval skeletal Sample from Klostermarienberg, Austria. *International Journal of Osteoarchaeology* 19, 695–710.

Van der Merwe, A.E., İşcan, M.Y., L'Abbé, E.N. 2006. The pattern of vertebral osteophyte development in a South African population. *International Journal of Osteoarchaeology* 16, 459–464.

Waldron, T. 1994. *Counting the Dead: The Epidemiology of Skeletal Populations*. John Wiley and Sons Ltd, Chichester.

Waldron, T. 2009. *Palaeopathology*. Cambridge University Press, Cambridge.

Waldron, T., Rogers, J. 1991. Inter-observer variation in coding osteoarthritis in human skeletal remains. *International Journal of Osteoarchaeology* 1, 49–56.

Weber, J., Czarnetzki, A., Spring, A. 2003. Paleopathological features of the cervical spine in the Early Middle Ages: A natural history of degenerative diseases. *Neurosurgery* 53, 1418–1424.

Weiss, E., Jurmain, R. 2007. Osteoarthritis revisited: A contemporary review of aetiology. *International Journal of Osteoarchaeology* 17, 437–450.

3. Enamel Hypoplasia in Post-Medieval London: A Reassessment of the Evidence for Childhood Health

Brenna R. Hassett

The aim of this paper is to provide a reassessment of the presence and prevalence of disruptions to growth during childhood in the form of dental enamel defects, specifically linear enamel hypoplasia (LEH). Some studies have suggested that defects are more prevalent in lower socio-economic status groups and are associated with reduced access to resources that increases the risk of growth-disrupting conditions, while other studies have shown higher prevalence in high-status groups. In this study, a comparison of methods, including microscopic surface tomographic analysis, for identifying enamel hypoplasia is presented and demonstrates that variation in methodology may contribute substantially to the lack of a consistent pattern in prevalence of LEH in past human populations. Unworn permanent teeth from five historically known assemblages of different socio-economic status from Post Medieval London provide a test case for demonstrating that the crude prevalence of LEH is insufficient to identify the scope of the experience of childhood health in the past. By different measures, either the highest status assemblages or the lowest can be shown to have a 'greater' experience of childhood growth disruptions, suggesting that identification and interpretation of defect occurrence is critical in assessing enamel hypoplasia prevalence.

Keywords: LEH; Linear enamel hypoplasia; Socioeconomic status; Dental Anthropology

1. Introduction

Enamel hypoplasia is a commonly recorded indication of growth disruption in human remains. Enamel hypoplasia forms part of a suite of indications of interruption to normal dental development termed developmental defects of enamel, or DDE, but it is the linear form of hypoplastic defect that is of special interest to studies of growth and development in archaeological contexts. Linear enamel hypoplasia, or LEH, is a defect arising when the normal process of enamel formation is interrupted by some systemic disturbance to the health of an individual, such as fever, malnutrition, or disease (Goodman and Rose, 1990). These defects follow the lines of regular development of enamel and so can be tied into the precise schedule of an individual's dental development (Hillson, 1992a). This allows the bioarchaeological or physical anthropological researcher the unique opportunity of accessing information on the health of an individual while they were living, rather than a simple 'snapshot' of health around the time of death. The vast majority of bioarchaeological standards call for LEH to be recorded (Buikstra and Ubelaker, 1994; Brickley and McKinley, 2004); this is done in order to understand the health of people in the past during the period of dental development – childhood.

Through comparative studies of childhood health, it is then possible to access differences in lived experiences by comparing groups based on different populations, socio-economic status, or sex. A great deal of research exists that outlines the prevalence of LEH in various human groups, both past and present. This research has been carried out on living subjects, archaeological remains, and a considerable variety of experimental animals (Pindborg, 1982; Goodman and Rose, 1991; Hillson, 1992a). There is strong consensus amongst researchers that defects represent interruptions to growth though the specific cause of a given disruption cannot be identified and LEH are

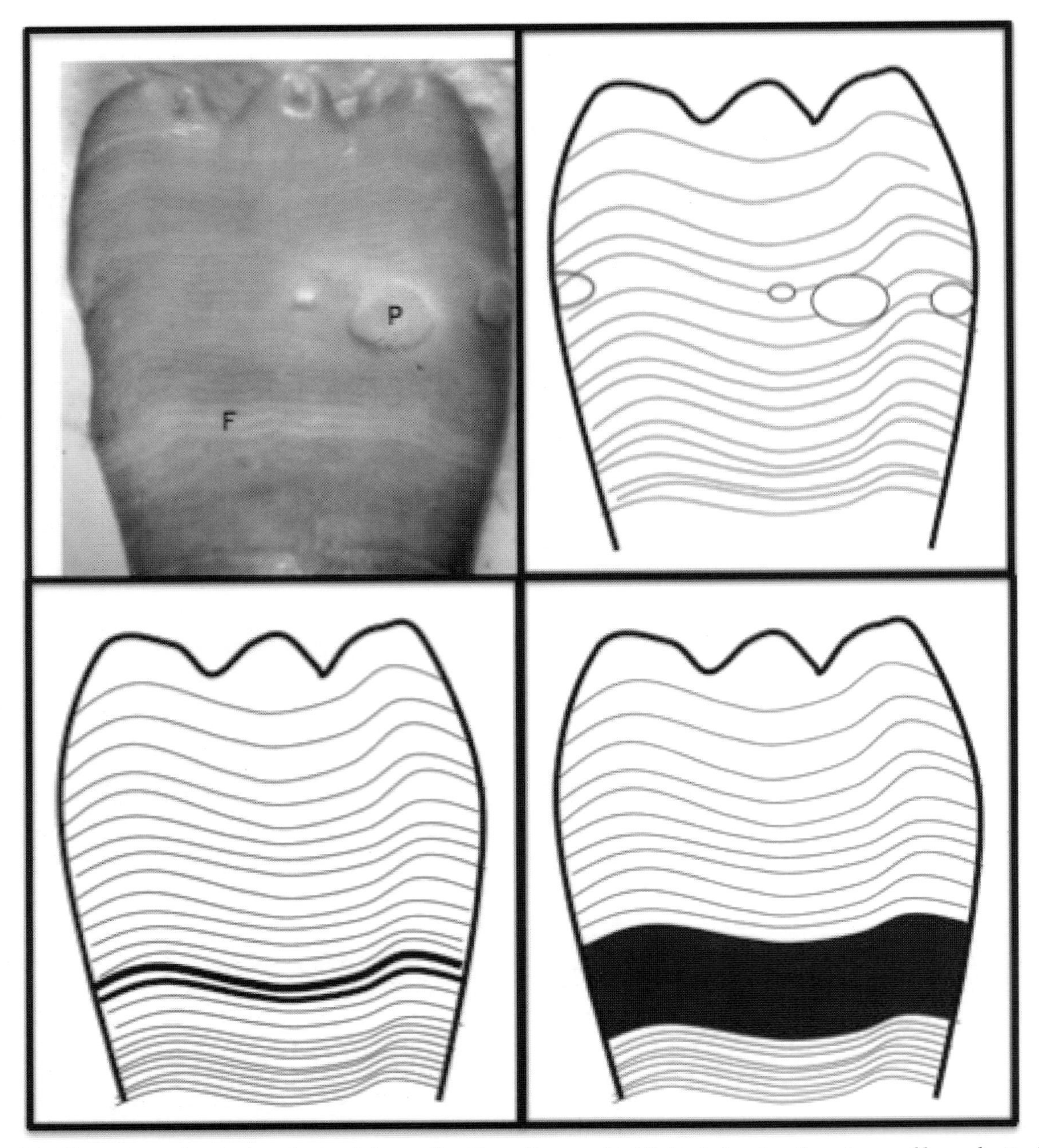

Figure 3.1. Illustration of an incisor (upper left panel) with perikymata highlighted showing three types of hypoplastic defect: pit form (upper right panel); linear form (lower left panel); and planar form (lower right panel). This study focuses on the linear form of defect (LEH).

best interpreted as generic markers of growth disruption (Hillson, 1992a). Where consensus breaks down is on the identification and interpretation of defect presence. Many studies of enamel hypoplasia do not set out criteria for assessing defect presence beyond the descriptive, with the main methods espoused in bioarchaeological standards calling for a visual assessment of the enamel surface supplemented by a tactile approach – generally, running a fingernail across the surface of the tooth to feel for a groove recorded (Buikstra and Ubelaker, 1994; Brickley and McKinley, 2004). Defect presence in these systems is defined by the observer. This becomes an issue when the form of the normal enamel surface is considered, however. The formation of enamel follows a regular developmental pattern, which is clearly reflected in the internal structures of the tooth (Boyde, 1989). Some of these regular structures are evident on the surface of the enamel as perikymata, circumferential 'wave-form' grooves and ridges that ring the tooth crown (Hillson, 1996). These perikymata form, depending on the individual, at roughly 7–11 day intervals, and are more closely spaced in the later-forming cervical parts of the tooth than in the earlier-forming cuspal areas.

Linear enamel hypoplastic defects are interruptions to the regular development of these structures, and can be described as areas where the perikymata distribution of the crown has altered – the even spacing between successive perikymata has been disrupted by areas of missing enamel. However, because the perikymata are further or closer apart depending on their location on the tooth crown, the 'size' of the defect will be determined by timing of the growth disruption. A defect occurring in a region of broadly-spaced perikymata will be metrically larger than one occurring at a time when closely-spaced perikymata are forming (Hillson, 1992a). Figure 3.1 shows an incisor with clear perikymata and multiple hypoplastic defect types.

Many studies do not distinguish between the three forms of hypoplastic defect – plane form, pit form, or linear (furrow) form – though it can be difficult to reliably assess the developmental timing of all but the linear form of defect. The issue of assessing developmental timing of a defect is critical to interpreting the experience of child health in the past, but there are extremely varied approaches based on multiple different schedules of dental growth used to estimate this, which can make cross-comparison between studies very difficult (Goodman and Rose, 1990). Methods range in resolution from the roughly 'eyeballed' assessment of defect occurrence in years, to the meticulous histological reconstruction of growth disruptions on a scale of days (Antoine *et al.*, 2005). These two factors, the discrepancy in identification of defect presence and the discrepancy in the interpretation of defect occurrence, may explain a large amount of the variation observed in LEH in the archaeological record (Hassett, 2011).

The present study considers methodological contributions to the variation in identifying and interpreting linear enamel hypoplasia. This is of critical importance considering the ubiquity of these analyses in bioarchaeology, and the value of the information LEH can provide. This comparison considers the effect of variation in *defect identification* by comparing different standards for assessing defect presence: macroscopic approaches ('naked-eye') versus microscopic approaches; definitions of presence which demand evidence of systemic growth disruption ('matched' defects, described further below) versus presence assessed through any evidence of growth disruption; and defects identified as present by the judgment of an observer versus defects identified using a metric approach. The effect of variation in *defect interpretation* is considered in terms of identifying the timing of defects using these variable approaches, depending on the resolution of data collected (defect timing assessed in years, months, days) and the overall interpretation of defect recurrence and periodicity.

2. Methods

Four methods of investigating LEH presence are presented here: one based on traditional field methods of macroscopic observation over broad timing categories ('Field Method') as described in the Institute of Field Archaeologists' Guidelines (Brickley and McKinley, 2004), one based on macroscopic observation of enamel but using a more exact schedule of dental development to categorize defects ('Macro' method), and two microscopic methods which observe defects at the level of the perikymata but are distinguished by using either observer judgment to define defect presence ('Micro' method) or a mathematical equation to do so ('Metric' method). A final consideration is whether it is necessary to distinguish between disruptions to growth evident on a single tooth from those that are evident on multiple concurrently forming teeth. If a hypoplastic defect is considered to be evidence of a systemic disruption to the growth of an individual in childhood, then there should be a systemic effect – all enamel that is forming at the time of the disruption should be affected and defects should 'match' across the dentition. Very few studies actually use this 'matching' criterion for identifying evidence of childhood health (exceptions include King *et al.*, 2002, 2005), and the results of introducing it are examined below.

There are four methods employed in this comparison, and three different methods of estimating age at defect formation. However, all the timing estimations are based on the histologically-informed enamel development schedules produced by Reid and Dean (2000, 2006) for permanent incisors, canines, and molars, and by Reid *et al.* (2008) for premolars. These schedules are broadly comparable to those produced by Schour and Massler (1941) that are traditionally applied in dental anthropology, but more accurately reflect the multiple rates of growth occurring in different tooth types over the course of childhood.

2.1 'Field' Method

Defect presence was identified in the Field method using the standards set out by the Institute of Field Archaeologists and the British Association of Biological Anthropologists and Osteoarchaeologists, which are commonly used in UK bioarchaeological research. The data was collected by experienced staff at the Museum of London Centre for Bioarchaeology and made available using the excellent Wellcome Trust funded WORD database which is freely available to the public (WORD Database, 2009). The standards of data collection are clearly stated in a downloadable document (Brickley and McKinley, 2004), also available from their website (http://www.museumoflondon.org.uk/Collections-Research/LAARC/Centre-for-Human-Bioarchaeology/Home.htm). This method identifies defects macroscopically, using the naked eye or low magnification (up to about 10×) to identify linear grooves or pitting on the enamel surface. Defect presence is assessed as either present or absent, based on observer judgement, and recorded for each third of the enamel crown. Unfortunately, time constraints prevented complete data collection for all materials by the museum, so there may be additional defects that have not been recorded using this method. Where possible, individuals who do not have any information on defect presence have been excluded from this study, but the possibility remains that there may be a slight undercount in the results for this method.

2.2 'Macro' Method

The Macro method identifies defect presence using macroscopic observation of the enamel surface with no, or low (10×) magnification. The surface of the enamel was examined by the author, and defects were recorded as present or absent based on observer judgement. The Macro method varies from the Field method in that it uses a greater degree of accuracy in estimating the timing of LEH by recording each defect according to tenths of enamel using a colour-coded graphic user interface on the front end of a MS Access database. This ensures that timing is estimated within the 2–3 month range of error associated with the Reid and Dean dental development schedules upon which it is based (Reid and Dean, 2000, 2006).

2.3 'Micro' Method

The Micro method identifies defects using microscopic observation of the enamel surface. This follows a method developed by Hillson and Jones (1989) and used in several subsequent studies to examine LEH at the level of the individual perikymata (Hillson, 1992a; Hillson and Bond, 1997; King *et al*., 2005). In this method, an epoxy cast is made of the enamel surface from a high resolution dental moulding material, then lightly coated in gold for visibility (Hillson, 1992b). This replica of the tooth surface is then examined under magnification of 100x, and where a hypoplastic defect is judged to be present by the observer, a note of which perikymata are affected is made. This allows the timing of the defects to be estimated in a much more exact manner, using the regular appearance of the perikymata to construct a roughly weekly record of development.

2.4 'Metric' Method

The Metric method is an adaptation of the Micro method, with the very important caveat that identification of defect presence is not based on observer judgement but instead on a quantitative approach which identifies significant alterations to the pattern of perikymata distribution (described in more detail in Hassett, 2012). This method accounts for the variable spacing of perikymata by constructing a local average of values, and identifies outlying values using a modified 'z-score' formula. As previously published, this can be written:

$$z = \frac{\chi - \mu}{\sigma}$$

"where χ is the value to be standardised, μ is the mean of all values, and σ the standard deviation of all values. In the modified z-score used here, χ represents the distance between two perikymata, μ is the local moving average of the focal measurement plus the 5 preceding and 5 succeeding distances, and σ is the standard deviation for the same set of measurements." (Hassett, 2012: 562)

A traditional z-score identifies values within a group that lie outside some boundary of statistical significance; the modified z-score used here is based on a moving average of standard deviations between observed and expected perikymata distances. In this case, areas where perikymata were outside 1.5 standard deviations of the normal distribution were considered as potential hypoplastic defects. This cut-off point was chosen using visual identification of defects as a guide, but can easily be adapted to reflect different research aims. Timing for defects was estimated in the same manner as for the Micro method.

3. Materials

The materials used for this study are permanent teeth drawn from five post-medieval London assemblages. They are taken from burials dating from around 1550–1850 AD, and represent a variety of socioeconomic strata of the burgeoning early modern city. All of the dentitions come from individuals who died between the ages of approximately 5 to 18 years of age. This age range was selected in order to provide the maximum number of complete, unworn enamel crowns for study. This deliberate age selection would make interpreting these results as a indication of mortality or survivorship in this group problematic but is carried out here because the subject of interest is understanding variation between observation methods, and the unworn teeth of this age group provides the maximum possible amount of data. 74 individuals in total were included in this study. Fourteen are drawn from the Lower Churchyard of St Bride's (site code FAO90), Fleet Street, a lower-status burying ground in a once-prosperous city parish that potentially holds the remains of persons dying in the notorious Bridewell and Fleet Prisons (Miles and Conheeney, 2005). Representing the more prosperous parishioners of the same church are eight individuals from the crypt burials, given the code 'SB' here (Bowman, 1997). Also included are 15 burials from the former grounds of St Benet Sherehog parish church (site code ONE94), dating from after the parish was amalgamated into that of neighbouring St Stephen Walbrook, representing the relatively comfortable London middle class (Miles *et al*., 2008). Another six burials come from the affluent parish church of Chelsea (site code OCU00), known as either All Saints or Chelsea Old Church (Cowie *et al*., 2007). Finally, 31 individuals are drawn from the large series of inhumations in the New Churchyard, one of the major non-consecrated burial grounds of the early modern city and populated largely by very low socioeconomic status individuals. The New Churchyard was used as an overflow burial ground for overtaxed city parishes during periods of epidemic mortality, to bury the destitute or foreign, and also chosen as a final resting place by prominent dissenting religious groups who did not wish to be interred under the state ecclesiastic system (Harding, 2000).

4. Results

There is considerable variation between the findings of the four different methods. One initial difficulty is that timing is assessed for several methods on different scales; to be able to directly compare the results means that instead of using direct counts of defective perikymata observed, or defects seen with the naked eye, it is necessary to substitute broad timing categories (half-yearly intervals) and note, for each tooth and each individual, whether or not a defect occurs in each of these intervals. The results below are therefore broadly comparable across methods, but, in the case of the microscopic approaches, lose considerable detail with regard to the timing of defects.

Table 3.1. Observed prevalence of LEH at the site of St Bride's Lower Churchyard (FAO90) across all four methods.

St Bride's Lower Churchyard	Field	Macro	Micro	Metric
Unmatched	18	34	111	160
Matched	6	9	30	94

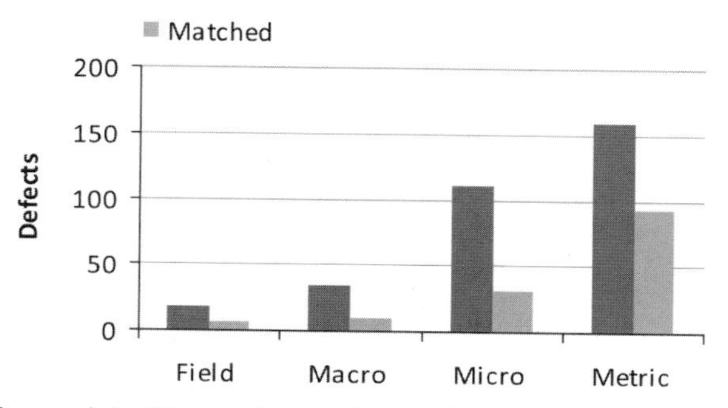

Figure 3.2. Observed prevalence of LEH at the site of St Bride's Lower Churchyard (FAO90) across all four methods.

One of the first areas of variation to address is the definition of defect presence, and whether defects are considered present if they occur in only one tooth in an individual; that is, they do not 'match' across simultaneously forming areas of the dentition. Table 3.1 presents, for just one site, the considerable difference between results using a 'matched' or 'unmatched' criterion to define defect presence, this data is presented graphically in Figure 3.2.

There are several different ways in which presence can be categorized for the five sites. Listing sites in order of number of defects will not necessarily match a list ordered on the number of individuals with defects, or amount of enamel affected overall. When overall LEH presence for a particular site is measured as the number of individuals with at least one defect, the results of the methods vary widely (Table 3.2, Figure 3.3).

Data collection by the Museum of London for the site of St Bride's Crypt and the New Churchyard was not complete at the time of publication, but a clear trend can be seen in Table 3.2. Much greater percentages of individuals are

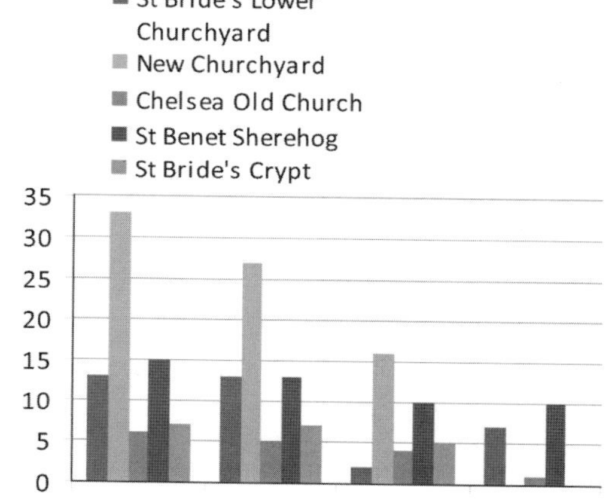

Figure 3.3. The number of individuals with at least one defect, as observed using all four methods.

Table 3.2. Percentage of individuals with at least one defect, as observed using all four methods.

Site	n	Field with LEH	Field %	Macro with LEH	Macro %	Micro with LEH	Micro %	Metric with LEH	Metric %
St Bride's Lower Churchyard	14	7	50.00%	2	14.29%	2	14.29%	13	92.86%
New Churchyard	33	n/a	n/a	16	48.48%	16	48.48%	33	100.00%
Chelsea Old Church	6	1	16.67%	4	66.67%	4	66.67%	6	100.00%
St Benet Sherehog	16	10	62.50%	10	62.50%	10	62.50%	15	93.75%
St Bride's Crypt	8	n/a	n/a	5	62.5%	5	62.5%	7	87.5%

Table 3.3. Comparative ranking of sites with the greatest percentage of enamel affected by LEH using four methods.

Highest % Enamel Hypoplastic (matched)	Field	Macro	Micro	Metric
1	St Benet Sherehog	St Benet Sherehog	Chelsea Old Church	St Bride's Lower Churchyard
2	Chelsea Old Church	St Bride's Crypt	New Churchyard	Chelsea Old Church
3	--	New Churchyard	St Bride's Lower Churchyard	New Churchyard
4	St Bride's Lower Churchyard	St Bride's Lower Churchyard	St Bride's Crypt	St Benet Sherehog
5	- -	Chelsea Old Church	St Benet Sherehog	St Bride's Crypt

Table 3.4. Average age of estimated defect formation across four methods.

Avg. Est. formation (years)	Field	Macro	Micro	Metric
Unmatched	3.2	2.9	3.0	3.6
Matched	2.9	2.7	3.2	3.5

seen to have some growth disruption using the microscopic methods (Micro and Metric). This significantly alters the interpretation of which site has the greatest number of individuals.

Using a microscopic approach to identifying defects will, invariably, turn up more defects. Defects which are not obvious at 10× may be perfectly obvious at 100×, and if observations are made for each perikymata, as they are here, then there are several hundred more opportunities to count a defect as 'present' than there would be in methods which are segmented into 3 or 10 observations (i.e. the Field and Macro methods, respectively). This can be clearly seen from the disparity in individuals with observed defects in Table 3.2, or in the total number of defects observed for St Bride's Lower Churchyard (Table 3.1). The most directly comparable approach is to calculate, for each method, the number of defects observed as a percentage of the total possible opportunities for observing a defect; this is shown in Table 3.3 by ranking the sites in order of the greatest amount of enamel affected.

A final consideration is the way in which methodological variation affects the interpretation of individual defects. Because the Metric method cannot use the first five perikymata observed, there is a strong likelihood that it undercounts defects occurring at the cusp of the tooth – these are the earlier forming defects. Conversely, the methods that rely on observer judgement to define defect presence are less able to identify disruptions in the more densely packed perikymata – which are concentrated in later-forming areas of the dentition. This is shown in Table 3.4.

5. Discussion

The remains of children buried in five post-medieval London burial grounds show no clear trend in LEH prevalence across different measures and methods of assessing growth disruption from LEH. The lowest status groups – those from the New Churchyard, and St Bride's Lower Churchyard – might be expected to show the greatest amount of evidence of childhood growth disruption. However, only the Metric method shows the two lowest status assemblages as having greater numbers of enamel affected, though all methods for which there are data agree that the New Churchyard has the highest number of individuals affected (Tables 3.2 and 3.3).

Introducing a matching criterion for defect presence has a clear effect on identifying defect presence. While the Metric method has the most consistent results, only 60% of defects observed occur in more than one simultaneously developing tooth. It is beyond the scope of this paper to comment on the potential impact of factors such as tooth morphology, enamel growth rate, aspects of the ameloblast life cycle, and severity of growth disruption on the formation of defects in different parts of the dentition; however it is clear from this study that there is a great deal still to be understood about the response of enamel to growth disruption. The greater consistency in results from the Metric method may indicate that one of the reasons defects are not seen to 'match' using other methods of observation is that small differences in scale are simply impossible to identify visually.

The variable results from multiple methods clearly demonstrate that the way in which defects are identified affects our understanding of the experience of childhood health in a given individual or site. Macroscopic methods clearly identify fewer defects. The introduction of a quantified criterion for defect presence also affects interpretation, with later forming defects more easily identified. Macroscopic methods seem to 'miss' defects that can be interpreted as evidence of systemic disruption to growth that occur in later childhood. This has a number of implications for studies that consider the average age of defect timing, particularly in those that relate 'peak' defect presence to socio-cultural practices such as weaning

(Corruccini *et al.*, 1985; Moggi-Cecchi *et al.*, 1993, 1994; Blakey *et al.*, 1994; Katzenberg *et al.*, 1996; Wood, 1996; Santos and Coimbra, 1999; Saunders and Keenleyside, 1999).

The variation observed in considering relative prevalence of LEH among groups of different socio-economic status can also be considered in terms of the timing and morphology of defects. Particularly in the case of the Chelsea Old Church sample, many individuals were observed to have defects occurring at the very cusp of the tooth. As mentioned above, it is not possible for a quantified method to pick this up. This suggests that a combination of methods may be needed to obtain the 'maximum' number of defects. Where the human eye is unable to successfully pick out defective enamel, a quantified metric approach may be applied; conversely, where the metric method cannot identify variation (at the very beginning and end of each tooth) it is only sensible to include the evidence of visual assessment of defect presence.

6. Conclusions

The choice of a particular scale of observation, and a particular method for defining defect presence, has a considerable effect on our understanding of childhood health in the past. While the results here show that a quantified, microscopic approach finds the greatest number of defects, it is almost certain that this is not a practicable approach for the majority of bioarchaeological investigations given that microscopic methods are particularly time intensive and the type of microscopic equipment needed impractical to carry into most field contexts. It is perhaps more useful to consider the results here as indicating that traditional field methods and other macroscopic approaches to identifying defects may 'miss' defects that occur in certain areas of the dentition (particularly the later-forming enamel). The variation in results when attempting to interpret the experience of childhood health in different socio-economic groups in post-medieval London suggests that it is important to clarify in future studies where method might affect interpretation of LEH. Hypoplastic defects impart critical information about health in the past but different levels of accuracy in timing estimation, defect definition, and ability to detect defects may all affect any final interpretation.

It is hoped that future work with both macroscopic and microscopic observation of defects will outline more clearly the discrepancies between different methodological approaches, and perhaps delineate some areas of the enamel where results are more consistent among different methods of observation and defect definition. There are a number of promising developments in the field of tomographic imaging, photogrammetry, and image analysis software which should also inspire a note of optimism for research into childhood health; with rapid improvement in digital imaging techniques it may be possible in future to gain highest-resolution information akin to that of the microscopic quantified approach examined here with far less cost to the researcher.

Acknowledgements

The author would like to thank S. Hillson and D. Antoine for their assistance and advice during the supervision of the PhD thesis from which this study is drawn, and to the organizers of the BABAO 2011 annual conference for arranging such an interesting and effective session. Additional thanks are due to A. Bevan, E. Bocaege, S. Bond, L. Humphrey, and C. Rando for a wide variety of comments and practical assistance; particular thanks go to J. Beklavac, R. Redfern, and to the much-missed Bill White for their unstinting help with accessing and understanding the collection. This research was undertaken with the support of Wenner Gren Dissertation Fieldwork Grant #7718.

References

Antoine, D.M., Hillson, S.W., Keene, D., Dean, M.C., Milne, G. 2005. Using growth structures in teeth from victims of the Black Death to investigate the effects of the Great Famine (AD 1315–1317). *American Journal of Physical Anthropology Supplement* 125, 65.

Blakey, M.L., Leslie, T.E., Reidy, J.P. 1994. Frequency and chronological distribution of dental enamel hypoplasia in enslaved African Americans: A test of the weaning hypothesis. *American Journal of Physical Anthropology* 95, 371–383.

Bowman, J. 1997. Documented skeletal collection, in: Milne, G. *St Bride's Church London: Archaeological Research 1952–60 and 1992–5.* English Heritage, Swindon, 93–95.

Boyde, A. 1989. Enamel, in: Berkovitz, B.K.B., Boyde, A., Frank, R.M., Höhling, H.J., Moxham, B.J., Nalbandian, J., Tonge, C.H. (Eds.), *Teeth.* Springer Verlag, New York, Berlin and Heidelberg, pp. 309–473.

Brickley, M., McKinley, J.I. 2004. *Guidelines to the Standards for Recording Human Remains.* IFA Paper No. 7. BABAO, Southampton and Institute of Field Archaeologists, Reading.

Buikstra, J.E., Ubelaker, D.H. 1994. *Standards for Data Collection from Human Skeletal Remains. Proceedings of a Seminar at the Field Museum of Natural History.* Arkansas Archeological Survey Research Series No. 44, Fayetteville.

Corruccini, R.S., Handler, J.S., Jacobi, K.P. 1985. Chronological distribution of enamel hypoplasias and weaning in a Caribbean slave population. *Human Biology* 57, 699–711.

Cowie, R., Bekvalac, J., Kausmally, T. 2007. *Late 17th- to 19th-century Burial and earlier Occupation at All Saints, Chelsea Old Church, Royal Borough of Kensington and Chelsea.* Museum of London Archaeological Service, London.

Goodman, A.H., Rose, J.C. 1990. Assessment of systemic physiological perturbations from dental enamel hypoplasias and associated histological structures. *American Journal of Physical Anthropology* 33, 59–110.

Goodman, A.H., Rose, J.C. 1991. Dental enamel hypoplasias as indicators of nutritional status, in: Kelley, M.A., Larsen, C.S. (Eds.), *Advances in Dental Anthropology.* Wiley-Liss, New York, pp. 279–293.

Harding, V. 2000. Death in the city: Mortuary archaeology to 1800, in: Haynes, I., Sheldon, H., Hannigan, L. (Eds.), *London Under Ground: The Archaeology of a City.* Oxbow Books, Oxford, pp. 272–283.

Hassett, B. 2011. *Changing World, Changing Lives: Child Health and Enamel Hypoplasia in Post Medieval London.* Institute of Archaeology, University College London. PhD Thesis, London.

Hassett, B. 2012. Evaluating sources of variation in the identification of linear hypoplastic defects of enamel: a new quantified method. *Journal of Archaeological Science* 39, 560–565.

Hillson, S.W. 1992a. Dental enamel growth, perikymata and hypoplasia in ancient tooth crowns. *Journal of the Royal Medical Society* 85, 460–466.

Hillson, S.W. 1992b. Impression and replica methods for studying hypoplasia and perikymata on human tooth crown surfaces from archaeological sites. *International Journal of Osteoarchaeology* 2, 65–78.

Hillson, S.W. 1996. *Dental Anthropology.* Cambridge University Press, Cambridge.

Hillson, S.W., Bond, S. 1997. Relationship of enamel hypoplasia to the pattern of tooth crown growth: A discussion. *American Journal of Physical Anthropology* 104, 89–103.

Hillson, S.W., Jones, B.K. 1989. Instruments for measuring surface profiles: An application in the study of ancient human tooth crown surfaces. *Journal of Archaeological Science* 16, 95–105.

Katzenberg, M.A., Herring, D.A., Saunders, S. 1996. Weaning and infant mortality: Evaluating the skeletal evidence. *American Journal of Physical Anthropology* 101(S23), 177–199.

King, T., Hillson, S., Humphrey, L.T. 2002. A detailed study of enamel hypoplasia in a post-medieval adolescent of known age and sex. *Archives of Oral Biology* 47, 29–39.

King, T., Humphrey, L.T., Hillson, S.W. 2005. Linear enamel hypoplasias as indicators of systemic physiological stress: Evidence from two known age-at-death and sex populations from postmedieval London. *American Journal of Physical Anthropology* 128, 547–559.

Miles, A., Conheeney, J. 2005. *A Post-medieval Population from London: Excavations in the St Bride's Lower Churchyard 75–82 Farringdon Street, City of London, EC4.* MoLAS Studies Series (unpublished).

Miles, A., White, W., Tankard, D. 2008. *Burial at the Site of the Parish Church of St Benet Sherehog before and after the Great Fire.* Museum of London Archaeological Service, London.

Moggi-Cecchi, J., Crovella, S., Bari, A., Gonella, P. 1993. Enamel hypoplasias in a 19th century population from northern Italy. *Anthropologischer Anzeiger* 51(2), 123–129.

Moggi-Cecchi, J., Pacciani, E., Pinto-Cisternas, J. 1994. Enamel hypoplasia and age at weaning in 19th-century Florence, Italy. *American Journal of Physical Anthropology* 93(3), 299–306.

Pindborg, J.J. 1982. Aetiology of developmental enamel defects not related to fluorosis. *International Dental Journal* 32, 123–134.

Reid, D.J., Dean, M.C. 2000. Brief communication: The timing of linear hypoplasias on human anterior teeth. *American Journal of Physical Anthropology* 113, 135–139.

Reid, D.J., Dean, M.C. 2006. Variation in modern human enamel formation times. *Journal of Human Evolution* 50, 329–346.

Reid, D.J., Guatelli-Steinberg, D., Walton, P. 2008. Variation in modern human premolar enamel formation times: Implications for Neandertals. *Journal of Human Evolution* 54, 225–235.

Santos, R.V., Coimbra, C.E., Jr. 1999. Hardships of contact: enamel hypoplasias in Tupi-Monde Amerindians from the Brazilian Amazonia. *American Journal of Physical Anthropology* 109(1), 111–127.

Saunders, S., Keenleyside, A. 1999. Enamel hypoplasia in a Canadian historic sample. *American Journal of Human Biology* 11(4), 513–524.

Schour, I., Massler, M. 1941. The development of the human dentition. *Journal of the American Dental Association* 28, 1153–1160.

Wood, L. 1996. Frequency and chronological distribution of linear enamel hypoplasia in a North American colonial skeletal sample. *American Journal of Physical Anthropology* 100(2), 247–259.

WORD Database 2009. *Wellcome Osteological Record Database.* Museum of London Centre for Human Bioarchaeology.

4. Archaeoanthropology: How to Construct a Picture of the Past?

Géraldine Sachau-Carcel, Dominique Castex and Robert Vergnieux

The discovery in 2004 of an unexplored section in the Saints Peter and Marcellinus catacomb, Rome, led to the development of a new method to visualise mass graves where individuals appear to have been inhumed simultaneously with specific funerary practices. This central section of the catacombs displays a unique organisation and dates from the late 1st to the early 3rd century AD. To study these mass graves, and to determine the patterns represented by the chronology of these deposits, an original method was created employing three-dimensional modelling with a protocol especially adapted to the burial context. The 3D modelling of the remains was based on field documents, on drawings and 3D photogrammetric records. A 3D scene was obtained regrouping all the components of the studied graves together with the architecture of the graves. Based on this scene, an analysis of the relationship between skeletal remains and taphonomic events within the deposits was undertaken in order to suggest chronological patterns. The possibility of visualising either a skeleton or a body from all angles facilitated the discussion on the functioning of the graves. The results confirm the simultaneity of some deposits and the existence of phases, probably related to several mortality crises.

Keywords: Three-dimensional modelling; Bioarchaeology; Mass grave; Rome; Catacomb

The discovery in 2004 of a previously unexplored part of the catacomb of Saints Peter and Marcellinus (Rome, Italy) has raised many questions. This part of the catacomb, known as the central sector, contains a large number of burials and was threatened by construction work necessitating its investigation. Beginning in 2004, several projects have been initiated in order to understand the nature and use of this sector.

1. Context

1.1 The Catacomb of Saints Peter and Marcellinus

The catacomb of Saints Peter and Marcellinus is located in the southeast of Rome, along the *via Casilina*, the former *via Labicana*. The catacomb has an area of three hectares and contains 20,000 to 25,000 individuals (Castex and Blanchard, 2009: 287). The graves are mostly individual, "loculi" (burial carved in the wall), or a regrouping of individual burials in "cubicula" (grouping several burials carved in the wall). However, the central sector discovered in 2004 is organised very differently to other parts of this catacomb and contains several cavities in which a large number of individuals were inhumed simultaneously (Figure 4.1).

The central sector dates from the end of the1st century to the beginning of the 3rd century AD. The funerary practices are unusual and most of the individuals are covered with textile and plaster, accompanied by amber, which gives them the appearance of mummies (Blanchard *et al.*, 2007: 990). In some cases, gold threads have been discovered completing the funeral apparatus. The preservation of the remains is variable and it is not possible to determine whether all individuals have undergone the same funerary treatment (Table 4.1, Figure 4.2).

Figure 4.1. Map of the catacomb of Saints Peter and Marcellinus and the central sector (PCAS, G. Sachau).

Figure 4.2. Photograph of a deposit, grave 16, scale: 0,40m (2006 season).

The biological and archaeological data studies suggest a major epidemic or a succession of mortality crises spread over time (Castex *et al.*, 2007, 2009, 2011). The hypothesis of a group of martyrs has been excluded because the skeletal remains do not present traumatic lesions.

Before the work to secure the site could be carried out, the remains (skeletal funerary apparatus) were studied and sampled for preservation. An Italian team[1] undertook the initial excavation, followed by a French team under the direction of Dominque Castex (DR, CNRS, UMR 5199) and Philippe Blanchard (INRAP, Centre). Four rescue excavations were organised and carried out in 2005, 2006, 2008 and 2010. These excavations estimated the number of individuals buried as approximately 3,000. Several university research projects have been launched in order to understand the type of population buried in this particular sector, to characterise the funerary practices and to discuss the mortality crises that led to the inhumation of a large number of individuals in a short time in this particular sector of the catacomb.

1.2 Aims and Objectives

To understand clearly the functioning of each grave classical methods are not optimal for the study of such a large number of individuals. Skeletal remains are present on several different field records. Sometimes there are significant differences in the depth at which anatomical sections

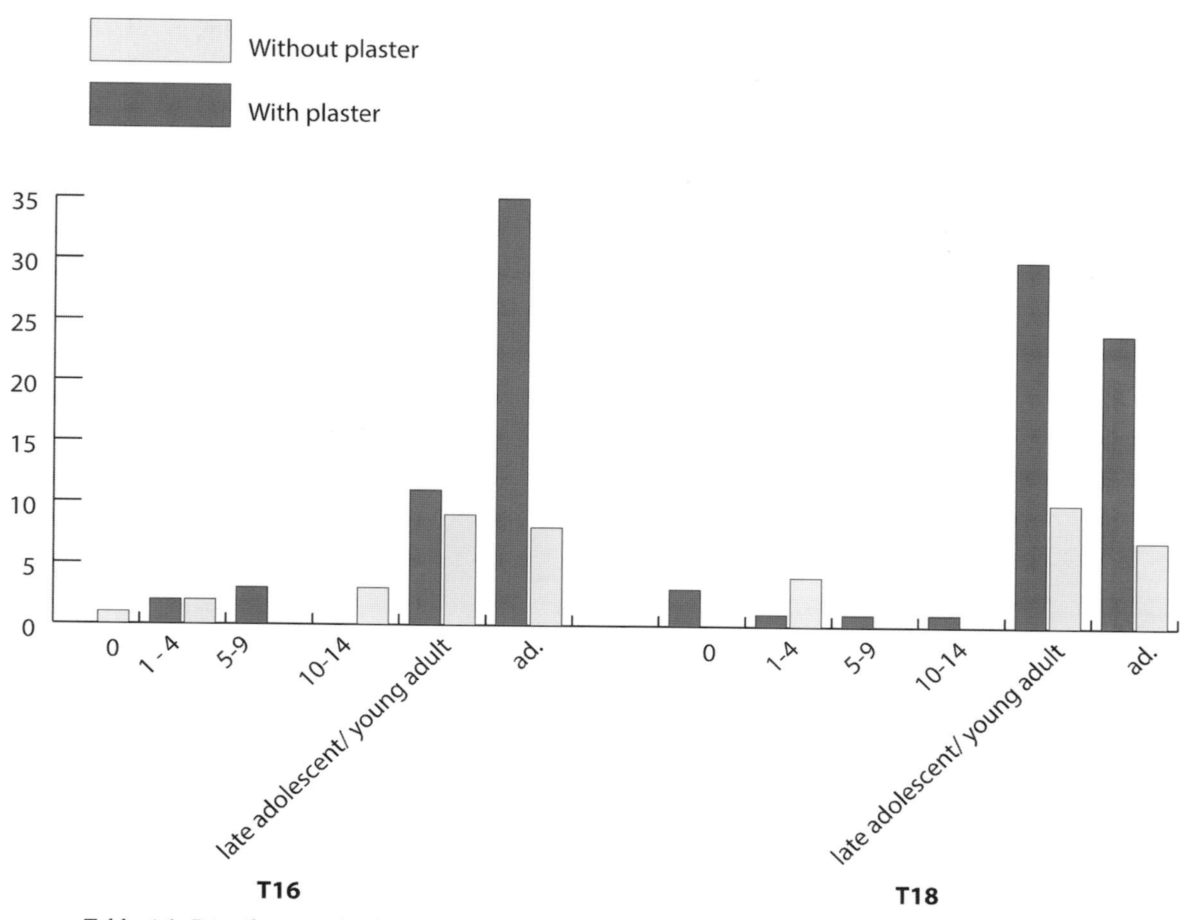

Table 4.1. Distribution of individuals by funerary treatment and age-at-death in graves 16 and 18.

from the same individual are found, and along with the state of preservation this does not permit the identification of all remains belonging to a particular individual. The relationship between individuals is particularly difficult to perceive as a result of these factors. In addition, the funeral practices employed negatively affect the study of the individuals, their inter-relationships and how the deposits were organised.

The characterisation of each individual deposit and of all deposits as a whole is essential to the discussion of whether these mass burials are the result of mortality crises. Furthermore, not all of the bodies seem to have been deposited at the same time, which is a central issue in the discussion of the mass burials. Given the singularity of the central sector and the unusual character of the deposits linked to the great number of individuals within each grave, it was extremely important to understand the functioning of these graves.

For this reason a three-dimensional modelling scheme was initiated by D. Castex in the form of a doctoral project.[2] In order to use this new method of representation and analysis, it was necessary to define the useful and necessary field documents to develop a protocol for 3D modelling.

1.3 Selection of Mass Graves

It was not possible to model all of the mass graves, so it was necessary to choose two multiple burials for which the excavation had been completed. These were fully excavated over two seasons in 2005 and 2006. These burials contained 78 and 76 individuals (T16 and T18, Sachau *et al.*, 2013). The methodology for the excavation and recording was that already used for other 14 multiple burials also excavated at Issoudun, a medieval site, by D. Castex and P. Blanchard (Blanchard *et al.*, 2011).

Grave 18 has an area of 7 m^2 and grave 16 of 2.50 m^2. At the time of the discovery, the combined thickness of the layers containing the human remains of 76 individuals, was 60 to 80 centimetres, which is an unusual phenomenon reflecting significant damage to the remains. The remains had suffered from differential preservation and the only evidence for some individuals was the "negative" traces left by the bones (in the sediment or in the funeral apparatus) (Figure 4.3).

The dominant position for the deposits was different for each grave, with decubitus positions predominating, but many cases of prone deposits and lateral positions were also observed (Table 4.2). The individuals were laid as primary deposits, with the labile articulations being in anatomical

Figure 4.3. Example of differential preservation in grave 18, scale: 0,40m (2006 season).

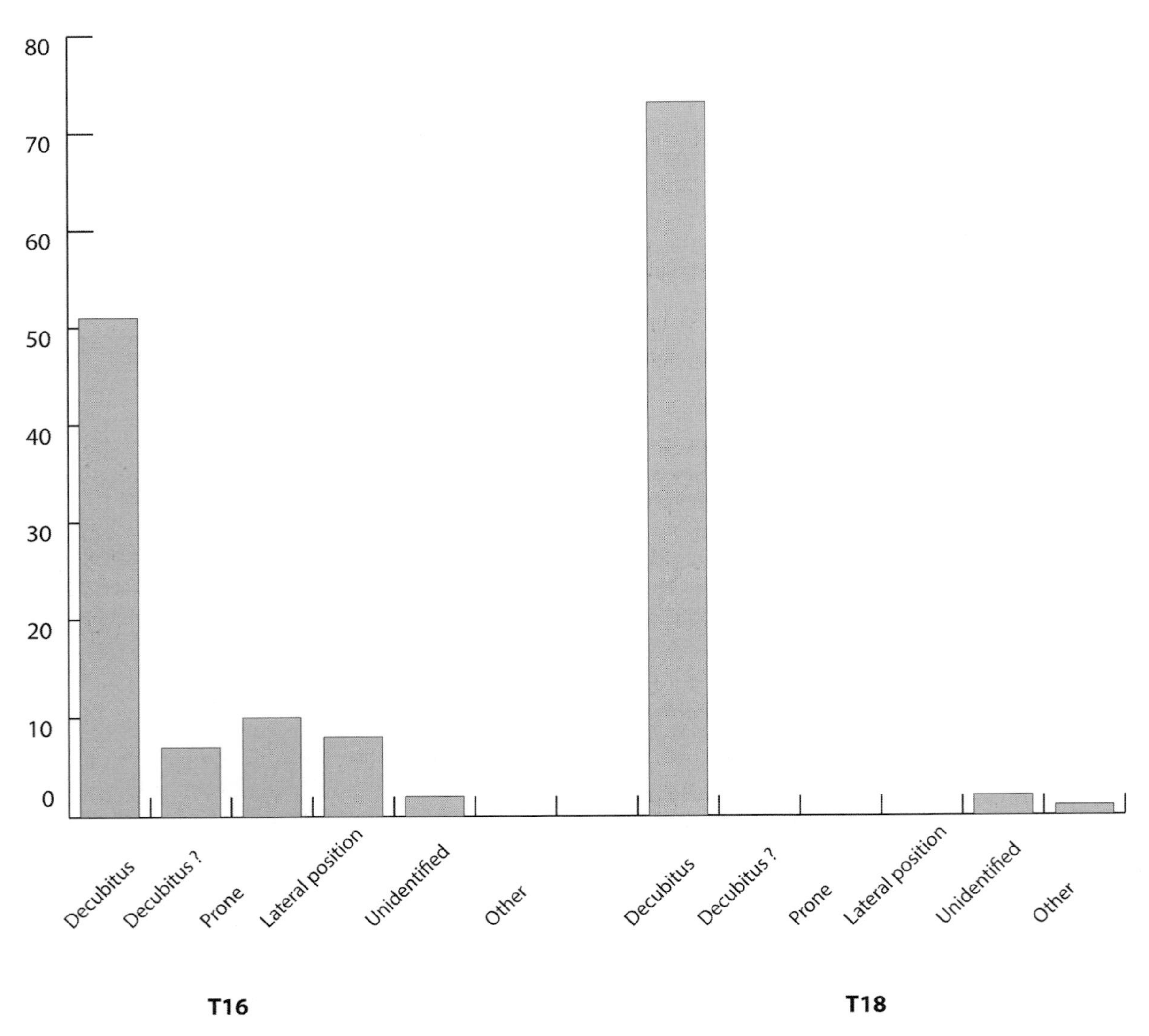

Table 4.2. Burial positions in graves 16 and 18.

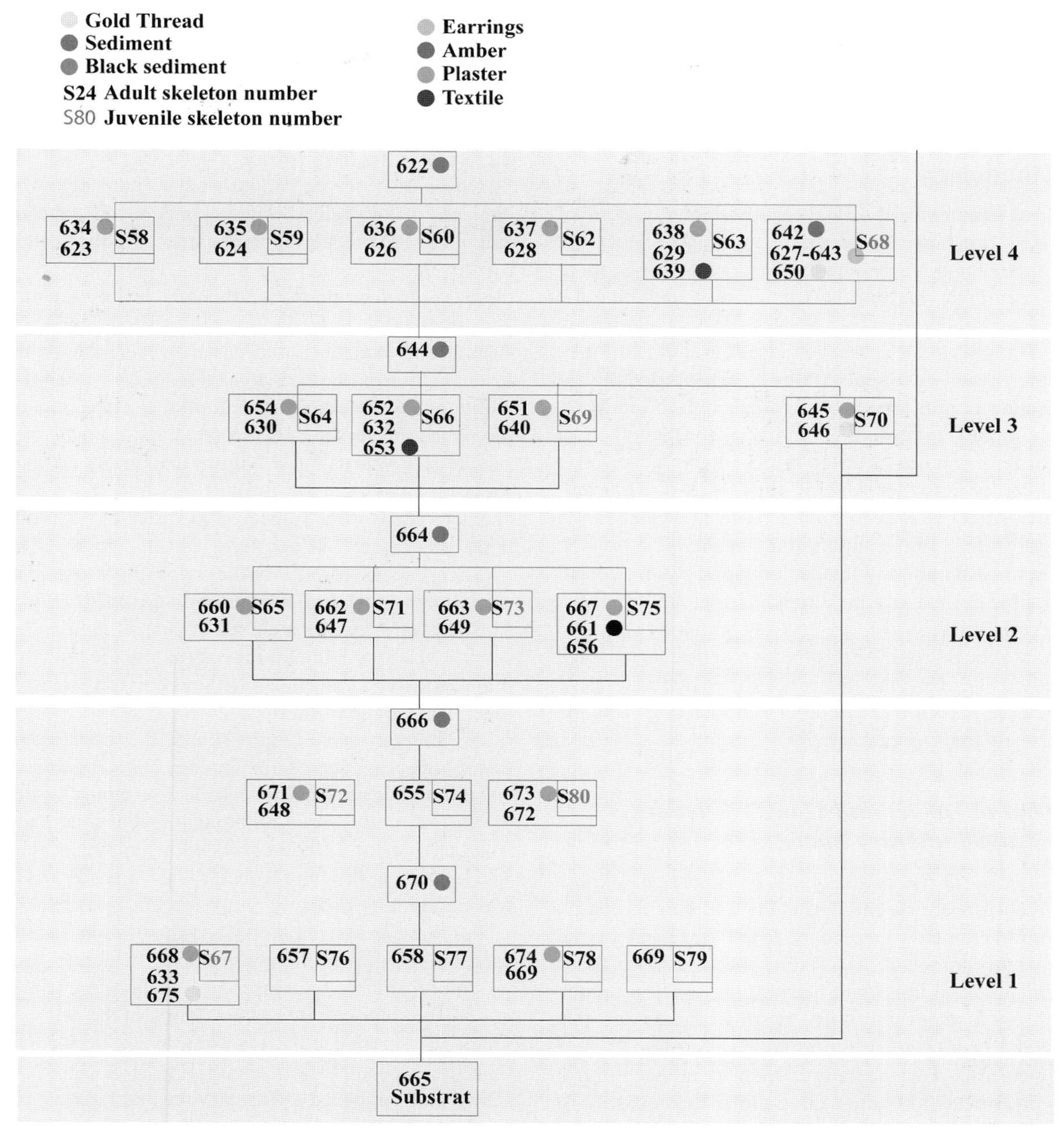

Figure 4.4. An extract from the stratigraphic record for grave 18.

connection when they were observable (Duday *et al.*, 1990). The displacements of the smallest bones were minor and linked to the decomposition of the body. However, the preservation of the remains was poor to mediocre concerning 74 to 89% of individuals. The quantification of the preservation is based on the macroscopic observation (aspect of the skeletal remains, friability, wholeness) and differed depending on the location in the grave, in particular along the walls due to water percolation. The taphonomic study was rendered more difficult due to the preservation of the remains and the funerary treatment they had undergone, especially those involving plaster.

The presence of levels of sediment of unequal thickness enabled the definition of strata of deposits. The graves were organised by levels of deposits, with 11 levels being observed for grave 18 and nine for grave 16. The number of individuals in each level was variable. Four to eight individuals had been deposited in the same level of grave 18, while in grave 16, level three contained twenty individuals (Figure 4.4). For grave 18, the sex of ten individuals has been determined and for grave 16 for only four individuals (Bruzek, 2002; Murail *et al.*, 2005).

It has not been possible to determine for either grave the time elapsed between each level of deposit, but only the overall chronology of the period of use. The organisation within each burial was particularly difficult to define. The large number of individuals deposited together with the complexity of the funerary treatment, makes the study and interpretation of the remains difficult.

2. Methodology

The three-dimensional (3D) visualisation of skeletal remains is not new. However, in most cases only the skeletal remains are represented, among other things, more or less accurately, but not the whole grave including the funerary architecture, as in 'Bodies 3D', developed by Richard Wright, and 'Crossbones' (Isaksen *et al.*, 2008). In the last ten years, new applications have proposed the 3D visualisation of one or more individuals, but the funerary space is merely indicated. In addition, the hypothetical (unrecorded) remains are not represented. It has been possible to reconstruct several sites with 'Laserscan', but the results have been aimed more at visualisation and interpretive presentation than analysis. Nor is it possible to easily employ this technique in a complicated stratigraphical context where remains are superimposed over several layers.

A new process was therefore developed that enabled the study of all elements present within a specific burial such as these mass graves, in order to understand, test and develop new lines of enquiry.

2.1 What to Depict and How? Skeletons, Bodies and Funerary Space

To produce a 3D reconstruction, it is important to know which elements are essential to study and thus to model, and which elements can be restored in terms of the archaeoanthropological issues. For these burials, all of the elements were important because the site is a unique example of a major mortality crisis in Rome for this period, and in a catacomb. The aim was to transform a two-dimensional representation, such as archaeological survey documentation, into a three-dimensional scene. To achieve this, it was necessary to place each bone and each element of funerary treatment and funerary space into a three dimensional frame of reference (Sachau *et al.*, 2010). The archaeological and anthropological excavation records were synthesised to determine what types of information were available and could be used for 3D modelling. The archaeological and anthropological data provided information about the sex, age-at-death and position of the individual, the surface of the bone that first appeared during the excavation (when possible), level of deposit and funerary treatments among other factors. For the analysis, it was necessary to consider how and in what form to depict this information in 3D.

For the graves already excavated the drawings provided the essential documentation as each detail observed in the grave was shown. In addition, the drawings enabled the spatial coordinates to be obtained, but were not sufficient to restore an "original picture of the past", that is, to retrieve the original layout of the grave and the bodies at the time they were deposited (Duday, 2007). For this reason, the construction of as complete a database as possible is an important step, every missing piece of information is as important as the information available.

It was also of vital importance to investigate how to depict first the skeletal remains and, second, the bodies, and which 3D models should be used. For the skeletons, a royalty free 3D model was chosen. The model adheres to the anatomical proportions although it is not a scientific 3D skeleton model. Indeed, prior to this project no complete scientific 3D skeleton model based on a reference collection or on a non-pathological skeleton was available. In addition, the skeletal remains from the central sector were too poorly preserved to be scanned and used as reference skeletons. It was useful to employ an entirely configurable skeleton in which the proportions and morphological characteristics were respected in order to position each bone as accurately as possible.

The process for the bodies was the same. The 3D model was produced by the software *MakeHumanTM*. This software enables a choice of size, weight and sex. It is particularly suitable for reconstructing the body in three dimensions and can be applied to individuals measuring over 1.10 metres in stature.[3]

The funerary space requires a specific treatment because this is always preserved and in many cases the dimensions and morphology of each grave have determined the deposits made. The central sector of the catacomb was composed of small cavities of different sizes with significant humidity. The lighting in the central sector was variable, and access was particularly difficult.

For these reason another 3D reconstitution method was used for the grave, photogrammetry,[4] which in this case was more suitable than laser scan technology (Sachau *et al.*, 2010). Photogrammetric recording of the entire central sector was carried out over a one-week period in 2010 with the Archeotransfert team (UPS SHS 3D N°3551 of CNRS and composed of Pascal Mora and Loïc Espinasse). The processing of the two graves studied in the current document was carried out following the recording, taking one week for each grave (Figure 4.5).

For the three-dimensional modelling of these multiple burials, it was decided to depict all of the remains found and recorded within the graves, to propose reconstruction hypotheses for the remains not present and to consider both skeletal remains and bodies. This working hypothesis was extremely important to study the individuals as they were deposited in the graves, and to understand the treatment of the bodies.

2.2 Creating a 3D Model from Field Documentation

The modelling of individuals was carried out in three steps. First, the 3D spatial coordinates were recorded.

Figure 4.5. An example of funerary space reconstruction using photogrammetry (P. Mora, G.Sachau).

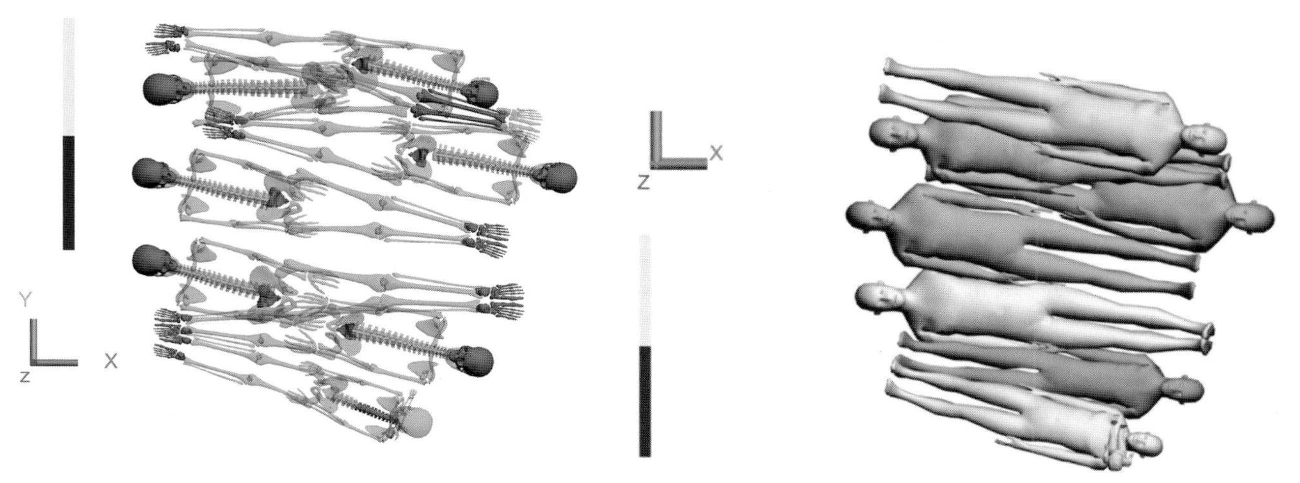

Figure 4.6. Steps in modelling deposit level 1 (3D skeleton and body), grave 18, scale: 1m (G.Sachau).

Secondly, the modelling of each element (bone, funerary treatment and funerary space) was carried out. Finally, all of the elements modelled were assembled in a single file, a 3D scene.[5] Individuals were depicted in three different forms: wireframe, 3D skeleton and body (Figure 4.6). Each step allows the visualisation in three dimensions of the interactions between each individual, the funerary treatments, the levels of sediment (which separate a level of deposit) and the funerary space.

The 3D coordinates for all remains were obtained from the archaeological records and converted into a vector drawing[6] following scale adaptation (Figure 4.7). The

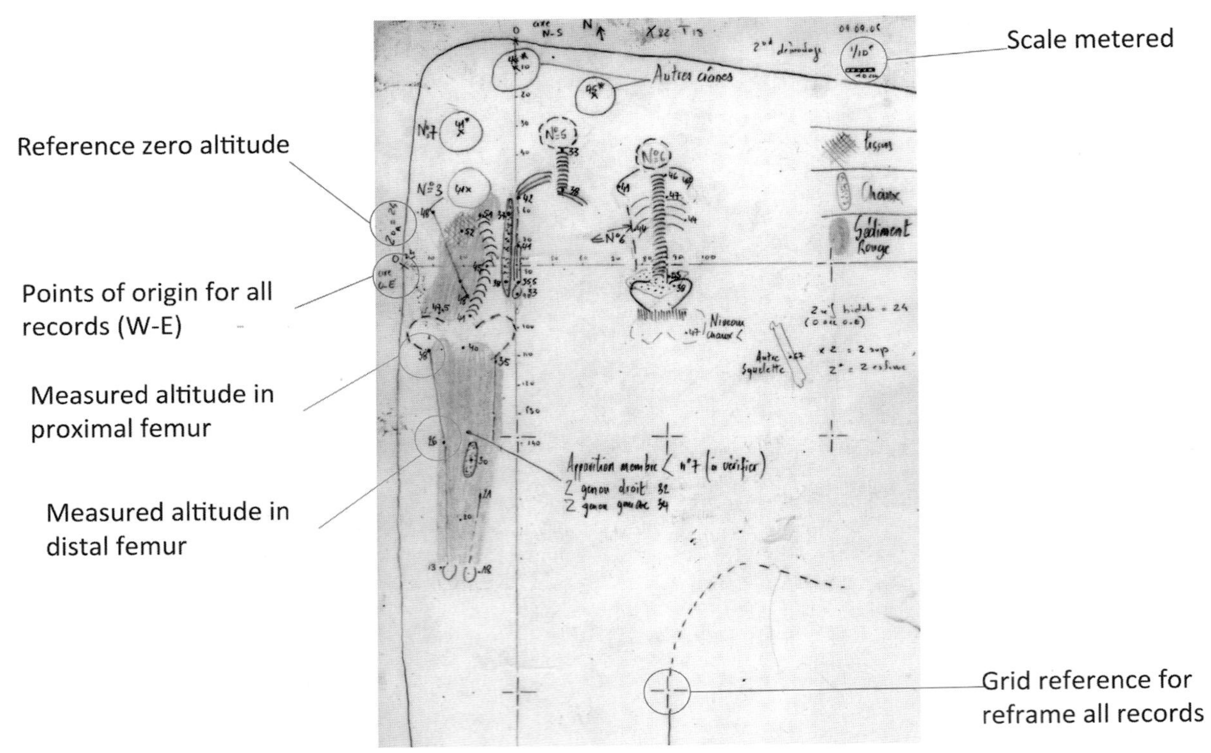

Figure 4.7. Spatial coordinate acquisition.

reference point(s) selected to determine the elevation[7] (depth) of each of the bones of an individual were defined and discussed before modelling with the archaeological and anthropological team. The upper elevation was set as two points for the long bones and one or two points for other elements such as the skull or os coxae, depending on the angle and preservation. For example, for the femur the proximal elevation was taken from the greater trochanter and the distal elevation was taken between the two condyles, as defined during the excavations. However, it was considered essential to know for each bone the surface (frontal, posterior, etc.) that had first been uncovered during excavation in order to identify accurately the point measured for the elevation. The information useful in reconstructing unrecorded remains did not appear on the drawings, but in the fields notes or in photographs, which is why it is of particular importance to construct as comprehensive a database as possible for each individual. For all remains, it was important to determine the X, Y and Z coordinates, but for the bones it was also necessary to know the elevation for each extremity (except for the skull and the sacrum), the observable surface during excavation, and other pieces of information such as sex and age-at-death. This information also helps consider the deposits and visualise the distribution according to sex, age-at-death or stature and more specifically to reconstruct the bodies. For the juvenile remains,[8] the point of elevation measured was also discussed with the team before modelling, and was located for the long bones at the extremities of the diaphysis; for the three-dimensional modelling, the same point was used.

All of the remains drawn were modelled; but a problem arose for the remains that were too small or too poorly preserved to be recorded, and which could not be modelled accurately, such as the gold threads. However, it was possible to position these approximately on the basis of the archaeological and anthropological observations made in the field. The procedure for the unrecorded bones was the same. It was not possible to draw some bones as they had not been preserved, or were only visible in negative; however, a number of conditions were defined that enabled these to be included in the 3D scene. The field documentation enabled the identification of the unrecorded bones, and it was then possible to work on the basis of the space left by the bone, as well as considering joint contiguity and symmetry (for the upper and lower limbs). The bones were depicted as opaque shapes if the bone was recorded and as transparent shapes if the position of the bone was estimated. The position of each bone was important in these multiple burials in order to understand the real position of the individual when deposited and the movements related to taphonomic phenomena.

During the modelling of the skeletal remains in three dimensions, questions arose as to how to represent the actual volume occupied by a skeleton if the only measurements available came from the surface of the bone which was observable during excavation. For the reconstruction, it was necessary to consider the full thickness or diameter of each bone and how the funerary treatment (plaster and tissues which cover some individuals) could have added to the volume of the actual body. For the funerary treatment it was impossible to measure the thickness, so an average thickness was determined depending on the samples. As a result of the state of preservation of the remains, the records did not take into account the thickness or diameter of the bones because it was impossible to measure them properly, the bones had undergone significant change and the thickness of the whole level containing the remains represented a maximum of 80 centimetres.

For the first level of deposit, in contact with the soil, it was possible to deduce the thickness of the bones, but it was also clear that the bones within both graves had undergone significant crushing and that the difference between the average real thickness/diameter and the observed thickness/diameter within the grave was considerable (for example, for the first level the pelvis measured approximately 4 centimetres antero-posteriorly, a particularly well preserved pelvis of an individual of the catacomb measured more than 15 centimetres). The solution chosen was to reproduce exactly the position of each bone, using the upper measurement[9] to place the 3D bone. This methodological choice led to another issue of non-anatomical interlocking, in which some individuals without apparent connection were not only in contact but also sometimes went through other individuals. To correct this type of error, all of the field documentation was studied in order to better understand the position of each of the bones and, by extension, each subject and to seek references to relationships between two or more interlocking subjects. If nothing had been observed, the relative position of each of the elements was calculated, taking into account their angle in order to decide how to position them.

For the reconstruction of bodies on the basis of the skeletal remains recorded, it was decided to represent the bodies as whole, and not only what was preserved. In this case a similar phenomenon was observed, with significant interlocking of bodies. In order to reconstruct the actual volume occupied by the bodies before decomposition, the assumption was made that no non-anatomical interlocking took place. The volume occupied by the body in the grave before decomposition had thus been depicted. The funerary material was represented as a geometrical shape because it was not always possible to delimit it accurately as a result of the material type.

The 3D modelling was tested first on grave 18, where the organisation of the individuals was simpler with little superimposition of remains, before being applied to grave 16. Once the model had been refined, the modelling of grave 16 required two weeks for the remains and one week for the funerary space.

3. Results

3.1 Archaeoanthropological Results of the Study of Mass Graves

The analysis of the 3D scene confirms the hypothesis formulated during the field study, indicating the

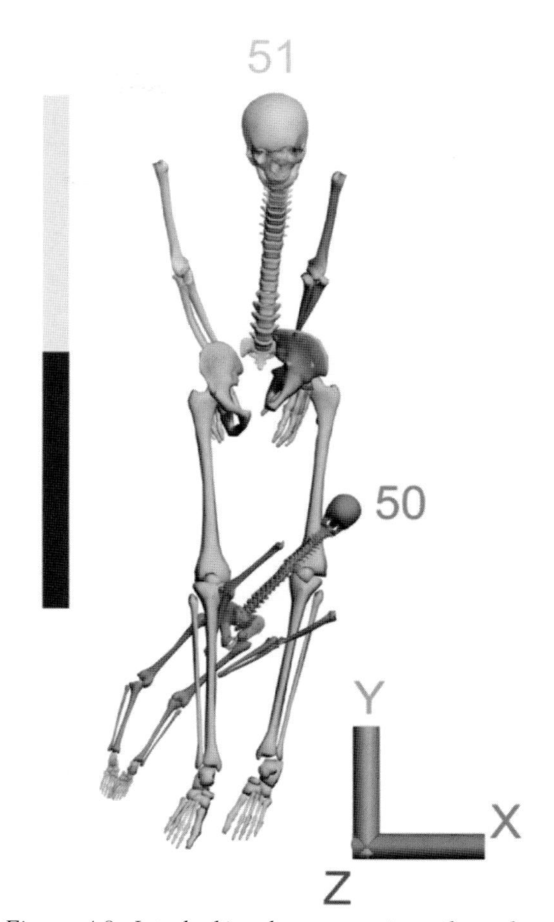

Figure 4.8. Interlocking between a juvenile and an adult, scale: 1m (G. Sachau).

simultaneous deposit of several individuals, highlighting the chronological phases involved and providing new elements for discussion on these graves as a result of the different levels of detail involved, from bone to the grave.

The most relevant and representative examples of the contribution made to the study of these complex graves by the 3D reconstruction will be discussed here.

For example, for an individual, the visualisation of all relationships with nearby individuals enabled his or her reallocation to a level of deposit, something which was not possible during the excavation due to the scarcity of remains.

The observation of these relationships also enabled the visualisation of the interlocking between two or more individuals. This interlocking was considerably more common in grave 16 than in grave 18, for an almost equivalent number of individuals deposited, approximately 80 individuals. In Figure 4.8, the interlocking between two subjects is shown, with the juvenile individual (50) overlapping the lower limb of an adult subject (51) and seemingly confirming simultaneity of deposition (Figure 4.8). Other examples were observed, discussed and approved by the archaeoanthropological team in view of the 3D reconstruction.

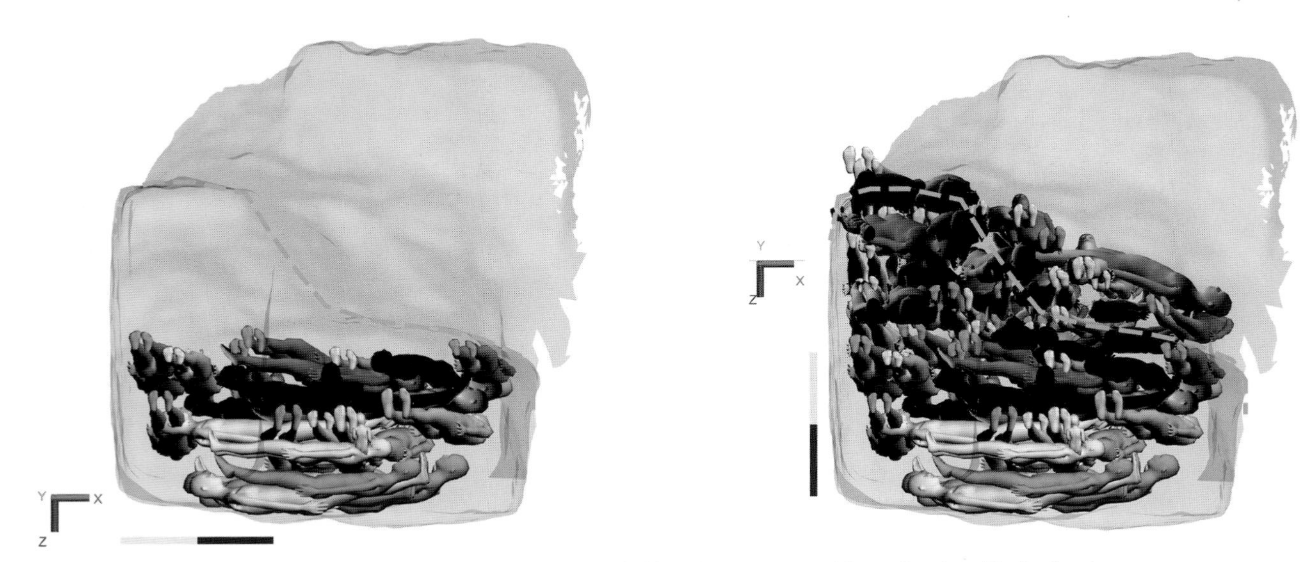

Figure 4.9. Frontal view (entrance) of all bodies in grave 18, scale: 1m (G. Sachau).

As a result of the preservation of the bones and the funerary treatment, the taphonomic processes within the two graves were complex. However, it was possible to observe the position of the individuals and the displacements and modifications in articulation, like disconnection between femora and tibia, relative position of the pelvic bones that ensued in particular due to the weight of other bodies. The modification of the position of the bodies' deposit due to the pressure exerted by the other bodies is an argument in favour of simultaneous decomposition. These observations constitute another point in discussing the chronology of the deposits, and several cases of simultaneous decomposition between levels are attested. Indeed, cases of simultaneous decomposition between some individuals within certain levels were demonstrated by the presence of "pull down effects" – this term is intended to define the position of a skeleton presenting a higher elevation of the skull and feet, relative to the pelvis, a position induced by the weight of the succeeding levels or by the free space released beneath the body as a result of decomposition and which does not correspond to the original deposit position of the body – in individuals found at different levels, which constitute an indication of the simultaneous decomposition of different layers of individuals. These effects are visible regardless of whether the individual was deposited on their back or front. For individuals in a supine position, the pressure exerted was so significant that the pelvis was located at a lower elevation than the skull and feet. In addition, for these individuals the lower limbs were found in a non-anatomical, hyper-extended or disconnected position. In grave 18, levels 7 to 11 were concerned and in grave 16, levels 1 to 7 presented the same effects. The study of the actual bodies complements such information, and the cumulated information enables the identification of chronological patterns linking all the available information.

Another type of information available from 3D modelling, which is not available through two dimensions is the estimation of the volume occupied by the bodies and then by the skeletonised remains in the grave. The 3D modelling of skeletons enables an assessment of the free space available after each deposit has decomposed and the 3D modelling of bodies makes it possible to visualise when the grave is full to determine chronological patterns for the use of the grave and examine the following two questions: (1) were all deposited at once, and (2) how many levels of deposits must decompose in order for it to be possible to deposit new bodies? With the non-interlocking 3D skeletons, the deposits reached ceiling height, but with the reconstruction of bodies, ceiling height was exceeded (Figure 4.9).

Therefore, for grave 18, only five consecutive levels could have been deposited. To deposit the last three levels (levels 8 to 11), the previous levels must have decomposed. For grave 16, the same observation can be made; ceiling height is achieved with six levels of deposit, but the entrance to the grave is blocked with five levels. To deposit the last three levels (levels 7 to 9), the six previous levels of bodies must have decomposed. Consequently, for both graves it was not possible for all bodies to have been deposited at the same time. The understanding of the use of the grave is based in part on arguments of simultaneity between individuals and levels, but also on waiting times for the decomposition of bodies. This information enables an understanding of the functioning of a grave during one or more successive mortality crises.

The 3D representation also enables the study of aspects such as circulation spaces through the visualisation of the free space following each level of deposit, distribution by sex, age-at-death, and the understanding of the logic behind the deposits and therefore the management of the bodies. A relative chronology has been proposed for each of the graves but it was not possible to accurately date each level

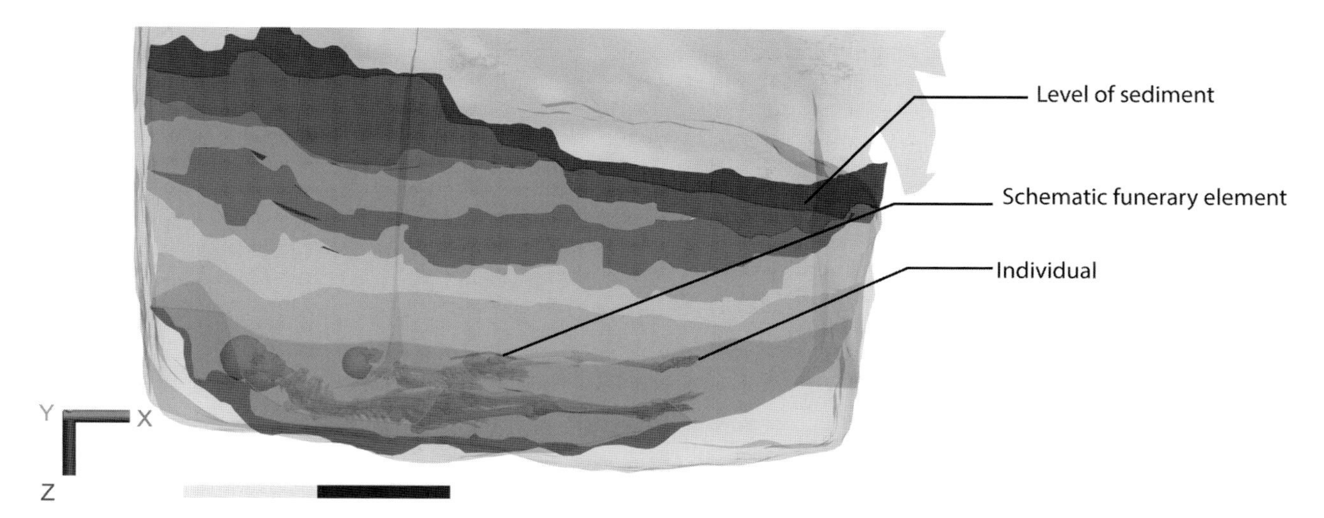

Figure 4.10. Frontal view (entrance) of the first level of deposit with funerary apparatus and the sediment levels, scale: 1m (G. Sachau).

either by conventional means as radiocarbon (too poorly preserved) or by the analysis of the model without other chronological patterns like funerary materials.

The analysis of the 3D scene for these two graves shows that their use was much more complicated than one might expect, with periods of simultaneous deposits and periods of waiting between deposits. These were thus mass graves with multiple sequential deposits.

3.2 What Other Contributions Compared to a Traditional Study?

The construction of 3D scenes allows the creation not of a single 3D picture but of an infinite number of 3D representations and the visualisation of multiple burials from every angle. The three-dimensional nature of the images also enables the accurate visualisation of all individuals at different levels from bone to skeleton to body and thus makes it possible to study the interlocking taking place between two or more subjects. The 3D scene is based only on the field documentation and synthesises it into a representation which ensures the reliability of the data. The image provided by the 3D scene, representing the remains exactly as recorded facilitates the discussion and stimulates interpretation.

All those working on the catacomb benefitted from discussing exactly the same depiction. More specifically, the study of the position of an individual was facilitated, because the 3D visualisation avoided the necessity to calculate the elevation for each of the bones or anatomical segments. It was therefore easier to compare the elevations at which the different bones of an individual were found, but also to compare the individual to other individuals in the same level. All analyses and hypotheses were discussed and approved by the researchers on the basis of the same visualisation which enabled the development of objective arguments in understanding the functioning of the central sector. In addition, the use of 3D models ensured the data was backed up, that the analyses were replicable and that the dataset was structured.

The 3D scene can evolve and be enhanced as a result of new hypotheses and new discoveries. The use of 3D tools allows different, more relevant results to be obtained than from a two-dimensional depiction, including hypothesis testing and a new form of visualisation.

The indication of chronological patterns is certainly the most important aspect of 3D modelling, with visualisation possible for each bone, all individuals deposited, the funerary treatments employed and the levels of sediment (Figure 4.10).

3.3 Other Applications? What Improvements are Required for Other Studies?

The other graves of the central sector are still being excavated and a comparison between graves may be of particular interest in understanding the overall functioning of this sector, during the mortality crises that took place in Rome from the late 1st to the early 3rd century AD.

This catacomb will perhaps never be opened to the public, but 3D reconstruction of the funerary space of the central sector already enables the visualisation of the variety and richness of the funerary complex represented by the catacomb. The 3D visualisation enables the presentation of the central sector and its multiple burials without it being necessary to visit it.

This type of representation is particularly adapted for the study of a large number of human remains in a complex stratigraphical context. Nevertheless, the creation of a skeleton, on the basis of a reference collection may improve the accuracy of the reconstruction and analysis. This project has enabled the development of a protocol and a methodology suitable for the construction of multiple burials and identification of the data essential to the modelling process, which can be applied to other sites. However, this protocol has yet to be tested for the modelling of unpublished multiple burials, those undergoing excavation or on a larger scale, in order to quantify the measurement errors and to identify more precisely the chain of inaccuracies. This work is naturally dependent on the techniques used, but also on the context of the research.

The prospects for the application of this modelling protocol are already significant with a possible extension to other multiple individual burials but also to collective burials. In addition, for individual burials, 3D reconstruction could help with visualisation and highlight archaeological assumptions such as the presence of a perishable container.

For collective burials, the ability to model all skeletal remains would be of use in the recognition of anatomical connections or remains in relative anatomical proximity. The visualisation of the distribution of the various anatomical segments of a single individual in the funerary space would also facilitate understanding and open up new paths of thought, as regards the treatment of individuals, whether as bodies or dry bones. The use of a digital scene in the study and comparison of a variety of graves could constitute a useful tool in understanding the funerary actions and practices carried out by the living population.

4. Conclusion

Seeking to reconstruct the original appearance of a grave is a challenge. The analysis of the 3D digital scene has enabled the identification of a set of results contributing to the understanding of multiples graves and complementing the data already collected relating to this particular sector of the catacomb of Saints Peter and Marcellinus, and confirming or refuting archaeological or anthropological hypotheses. In creating and applying this reconstruction protocol, it has been possible to identify the information which is essential to record in a burial context, such as the spatial coordinates and the observable face of the skeletal remains during excavation, and to propose a new form of recording in order to synthesise the biological and archaeological data (Sachau, 2012). In-situ recording could enable the study in the field of the position of the individual and the relationship between individuals and thus improve the comprehension of a burial with multiple deposits. 3D modelling offers a new method for the representation and visualisation of multiples burials which contributes to new research perspectives. The characteristics of 3D modelling enable the proposal of a complete tool for the study of burials. The use of 3D scenes in the representation of burials could result in the creation of real virtual libraries facilitating exchanges and comparisons relating to different sites.

The understanding of the choices made by those responsible for the burials is a recurring problem in the study of multiple graves and 3D reconstruction is an excellent tool for the analysis of sites with complex stratigraphical contexts, and particularly for the understanding of the management of the bodies.

Acknowledgements

Pontificale Commissione di Archeologia Sacra (Rome, Italy) and especially Raffaella Giuliani, Maison des Sciences de l'Homme d'Aquitaine (Bordeaux, France), UPS SHS 3D 3551 CNRS (Bordeaux, France).

Notes

1. This excavation was carried out under the direction of Raffaella Giuliani (Chief inspector of Rome catacombs) and Monicca Ricciardi (pontifical archaeologist).
2. PhD thesis "Apport de la modélisation tridimensionnelle à la compréhension du fonctionnement des sépultures multiples. L'exemple du secteur central de la catacombe des Saints Pierre-et-Marcellin (1er–IIIe ap. J.-C.) (Rome, Italy) defended at the University of Bordeaux 3, France, on 12 October 2012 by G. Sachau-Carcel.
3. Measurement of the femur in-situ or in the laboratory was used in this context to estimate the stature of each individual. Where this had not been preserved, the dimensions of other bones were compared in order to estimate the stature (see Cleuvenot and Houët, 1993: 245–255; Sachau, 2012).
4. Photogrammetric records were made with a Nikon 60D and the software used was PhotoModeler Scanner® and Geomagic®.
5. Developed with 3D Studio Max®.
6. Using Adobe Illustrator®.
7. Z coordinate.
8. 21 juveniles under 19 years were identified for graves 16 and 18.
9. Each point of measurement for each bone was defined with the archaeological and anthropological teams.

References

Blanchard, Ph., Castex, D., Coquerelle, M., Giuliani, R., Ricciardi, M. 2007. A mass grave from the catacomb of Saints Peter and Marcellinus in Rome, second-third century AD. *Antiquity* 81, 989–998.

Blanchard, Ph., Souquet-Leroy, I., Castex, D. 2011. Issoudun, Indre, Les Champs Elysées, (Centre de l'image), Témoignages de deux crises de mortalité moderne dans le grand cimetière, volume 1: texte et figure, volume 2: catalogue des sépultures individuelles, volume 3: catalogue et étude pathologique des sépultures multiples, volume 4: annexes: études documentaire et céramique, inventaires techniques, INRAP, unpublished excavation report.

Bruzek, J. 2002. A method for visual determination of sex, using the human hip bone. *American Journal of Physical Anthropology* 117, 157–168.

Castex, D., Blanchard, Ph., 2009. Les sépultures du secteur central de la catacombe des Saints Pierre-et-Marcellin (Rome). État des analyses bio-archéologiques et perspectives, *Mélanges de l'Ecole Française de Rome – Antiquité* 121(1), 287–297.

Castex, D., Blanchard, Ph., 2011. Le secteur central de la catacombe des Saints Pierre-et-Marcellin (Rome, Ier–IIIe siècle). Indices archéologiques d'une crise brutale de mortalité. *Mélanges de l'Ecole Française de Rome – Antiquité* 123(1), 274–280.

Castex, D., Blanchard, Ph., Giuliani, R., Ricciardi, R. 2007. Les ensembles funéraires du secteur central de la catacombe des Saints Pierre-et-Marcellin (Rome, Ier–IIIe siècle): caractérisation, hypothèses d'interprétation et perspectives de recherches. *Mélanges de l'Ecole Française de Rome – Antiquité* 119(1), 274–282.

Cleuvenot, E., Houët, F. 1993. Proposition de nouvelles équations d'estimation de stature applicables pour un sexe indéterminé, et basées sur les échantillons de Trotter et Gleser. *Bulletins et Mémoires de la Société d'Anthropologie de Paris* n.s. 5(1–2), 245–255.

Duday, H. 2007. Archaeological proof of an abrupt mortality crisis: Simultaneous deposit of cadavers, simultaneous deaths?, in: Raoult, D., Drancourt, M. (Eds.), *Paleomicrobiology and Past Infections*. Springer, Berlin and Heidelberg, pp. 49–54.

Duday, H., Courtaud, P., Crubezy, E., Sellier, P., Tillier, A.-M. 1990. L'Anthropologie «de terrain»: reconnaissance et interprétation des gestes funéraires, in: Crubézy, E., Duday, H., Sellier, P., Tillier, A.-M. (Eds.), Anthropologie et Archéologie: Dialogue sur les Ensembles Funéraires, Réunion de Bordeaux, 15–16 juin 1990. *Bulletins et Mémoires de la Société d'Anthropologie de Paris* n.s. 2(3), 29–49.

Isaksen, L., Loe, L., Saunders, M.K. 2008. *X-Bones: A New Approach to Recording Skeletons in 3D*. Poster presented at the British Association of Biological Anthropology and Osteoarchaeology Annual Conference, Oxford 2008.

Murail, P., Bruzek, J., Houët, F., Cunha, E. 2005. DSP a tool for probabilistic sex diagnosis using worldwide variability in of hip-bone measurements. *Bulletins et Mémoires de la Société d'Anthropologie de Paris* n.s. 17(3–4), pp. 167–176.

Sachau-Carcel, G. 2012. *Apport de la modélisation tridimensionnelle à la compréhension du fonctionnement des sépultures multiples. L'exemple du secteur central de la catacombe des Saints Pierre-et-Marcellin (Ier–IIIe ap. J.-C.) (Rome, Italie)*. Thèse de Doctorat, Université de Bordeaux 3.

Sachau, G., Castex, D., Mora, P., Vergnieux, R., 2010. Modélisation de deux ensembles funéraires de la catacombe des Saints Pierre-et-Marcellin à Rome: objectifs et méthodes, in: Vergnieux, R., Delevoie, C. (Eds.), *Actes du Colloque Virtual Retrospect 2009*. Archéovision 4, Éditions Ausonius, Bordeaux, pp. 161–170.

Sachau-Carcel, G., Vergnieux, R., Castex, D. 2013. Sites à stratification complexe et modélisation tridimensionnelle: Vers une nouvelle approche des sépultures multiples. *Archeosciences* 37, 89–104.

5. Palaeopathology of the Isle of May

Marlo Willows

The Isle of May is a small island off the east coast of Scotland. During the medieval period, the Isle of May was a pilgrimage site with a monastery dedicated to St. Ethernan. Excavations of this monastery were completed between 1992 and 1997 by the Glasgow University Archaeology Research Division. The excavations revealed fifty-eight articulated burials dating between 430 to 1580 CE. This paper evaluates the claim that the Isle of May had a healing tradition during the medieval period by re-analysing the health of the skeletons buried at the Isle of May and comparing it to another Scottish medieval community, from the site of Ballumbie. The significantly higher prevalence for disease as well as the variety of rare diseases found among the burials from the Monastery of St. Ethernan makes it probable that religious pilgrimage was not the only purpose for the island. Statistical analysis revealed individuals buried at the Isle of May had three times as many skeletal lesions than individuals buried at Ballumbie. The findings here strengthen the claim that the Isle of May likely had a healing tradition unrelated to the Benedictine priory; the majority of activity from the 8th to the 12th century.

Keywords: Medieval; Scotland; Ballumbie; St. Ethernan monastery; Skeletal lesions; Bioarchaeology

1. Introduction

The Isle of May is a small island in the Firth of Forth, 8 kilometres off the coast of Scotland. The island is currently owned and managed by Scottish Natural Heritage as a national nature reserve; boats regularly take visitors to bird watch from Anstruther, Scotland. The Isle of May St. Ethernan monastery site was excavated between 1992 and 1997 by Glasgow University Archaeological Research Division. The boundaries of the original monastery were discovered as well as the multiple building phases. A cemetery with both long cists and dug inhumations were exposed. The earliest radiocarbon date for the cemetery is 430 CE ± 70 years and the most recent date is 1580 CE ± 50 years (James and Yeoman, 2008). During the medieval period, the Isle of May was a very religious location, first probably housing Celtic wooden churches as early as the 5th–7th century; and later in the 12th century, it was established as a Benedictine priory (Duncan, 1956; Eggeling, 1960; James and Yeoman, 2008). The later church was dedicated to St. Ethernan, who is claimed to have died on the island in 669 CE. The priory on the Isle of May was founded as a dependant priory to Reading and was abandoned in favor of Pittenweem, an Augustan Priory on the east coast of Scotland, in the late 13th century (Dilworth, 1995; James and Yeoman, 2008). Although the priory re-located, the monks still maintained the church and land for another two centuries. Due to its location and religious significance, the Isle of May was a popular pilgrimage destination throughout the medieval period (Duncan, 1956, Eggeling, 1960, Yeoman, 1999; James and Yeoman, 2008). This paper will investigate the claim that the island was also known as a place of healing through re-analysis of the articulated remains and comparison to another Scottish medieval site in eastern Scotland, Ballumbie, in County Angus.

A descriptive report of the Isle of May burials has been published (James and Yeoman, 2008). Based on radiocarbon dates, the burials are organized into six burial groups specific to a period of occupation in which 48 of the individuals were buried between the 5th and mid-12th century; 10 individuals were buried between the 11th and 17th century (James and Yeoman, 2008). This indicates that most of the occupation and use of the cemetery was before the 13th century, corroborating what is known about the island historically.

Disease in the Middle Ages was often associated with sin; repentance and prayer was required for healing (Durkan, 1962; Daniell, 1997; Hamilton, 2003). There were three levels of health care: self-help, folk healers, and professional advice (Hamilton, 2003). Most individuals, rich or poor, knew enough about healing in order to dress minor wounds or fix simple medicines. Self-care also included using healing stones, visiting healing wells, and pilgrimage to religious sites (Hamilton, 2003). Most towns, even small ones, generally had a folk healer, usually thought to have a supernatural ability (Hamilton, 2003; Roberts and Cox, 2003). Folk healers could accomplish more; incorporating prayers or charms, mixing potions and tonics, and occasionally bloodletting. Professional advice was not available to most rural people, but in larger towns there were surgeons or barbers that could perform minor surgeries, set fractures, remove teeth, and perform bloodletting (Roberts and Cox, 2003). There were few hospitals in the modern sense in the Middle Ages; those that did exist were usually outside of larger towns and generally were founded to care for specific diseases such as leprosy, plague, and in the late middle ages, syphilis (Ewan, 1990; Hamilton, 2003; Roberts and Cox, 2003; Hall, 2006). Some hospitals, or almshouses, also cared for the poor and the elderly by feeding them and offering prayers for their souls (Daniell, 1997; Hall, 2006).

Although there is no known historical evidence that the Isle of May had a 'healing centre', the Aberdeen Breviary, written in 1509, claims that by drinking the water from the pilgrims' well "barren women, especially coming in the hope of thereby becoming fruitful, were not disappointed (cited in Stuart 1868)." The breviary continues to say the island was the "scene of many reputed miracles" (cited in Eggeling, 1960). Healing wells were not uncommon in medieval Scotland, many allegedly blessed by saints such as St. Columba and St. Ninian (Hamilton, 2003). As mentioned earlier, healing wells were a common part of health care. Examination of the environmental material on the Isle of May revealed two types of plants with medicinal properties used during the medieval period: *Hyoscyamus niger,* commonly known as black henbane, and *Chelidonium majus,* commonly known as greater celandine (James and Yeoman, 2008). The historical and environmental evidence support a healing tradition theory and the results of the re-evaluation of the skeletal remains strengthen this argument.

2. Materials and Methods

The skeletal remains from the Isle of May are curated by the National Museum of Scotland and they have been on loan to the University of Edinburgh since October 2010. While there are disarticulated remains in the assemblage, only the articulated burials will be discussed here for accuracy and consistency. The 58 articulated skeletons recovered represent the estimated twenty percent of the cemetery that was excavated between 1992 and 1997 by Glasgow University Archaeological Research Division. Preservation was evaluated by completeness and quality of the cortical bone. Table 5.2 illustrates that 39.7% of bone preservation was good, 50% was fair, and 10.3% was poor, as defined by guidelines recommended by Steven Byers (2010).

Methods used for age determination include: epiphyseal fusion (Schaefer *et al.*, 2009); dental development (Van Beek, 1983; Ubelaker, 1999); dental attrition (Brothwell, 1981); auricular surface morphology (Lovejoy *et al.*, 1985); pubic symphysis morphology (Brooks and Suchey, 1990); and changes to the sternal end of the 4th rib (İşcan *et al.*, 1984). For the purposes of this paper, 'subadult' refers to individuals under the age of 18 and 'young individuals' refers to individuals under the age of 25. Age categories are summarised in Table 5.3.

Table 5.1. Burial activity data compiled from James and Yeoman 2008.

Burial Group	Date	Number of Burials
1	5th–7th centuries	13
2	8th–mid 12th centuries	21
3	8th–11th centuries	14
4	11th–13th centuries	7
5	13th century	2
6	14th–17th centuries	1

Table 5.2. Bone preservation.

Preservation	Count	Percentage
Good	23	39.7
Fair	29	50.0
Poor	6	10.3
Total	58	100.0

Table 5.3. Age categories.

Neonate	0–2 mo
Infant	2 mo–2yr
Young Juvenile	2–6
Old Juvenile	7–12
Adolescent	13–17
Young Adult	18–24
Young Middle Adult	25–35
Middle Adult	36–45
Old Adult	46+
Adult	18+
Subadult	< 18
Young individual	< 25

Table 5.4. Isle of May demography (YJ=Young Juvenile, OJ=Old Juvenile, ADL=Adolescent, YA=Young Adult, MA=Middle Adult, OA=Old Adult).

			Sex				
			Male	Female	Undetermined	Non-Adults	Total
Age	YJ	Count	0	0	0	2	2
		% within Age	.0%	.0%	.0%	100.0%	100.0%
	OJ	Count	2	0	0	1	3
		% within Age	66.7%	.0%	.0%	33.3%	100.0%
	ADL	Count	8	0	0	0	8
		% within Age	100.0%	.0%	.0%	.0%	100.0%
	YA	Count	9	1	1	0	11
		% within Age	81.8%	9.1%	9.1%	.0%	100.0%
	MA	Count	14	1	1	0	16
		% within Age	87.5%	6.3%	6.3%	.0%	100.0%
	OA	Count	11	0	0	0	11
		% within Age	100.0%	.0%	.0%	.0%	100.0%
Total		Count	44	2	2	3	51
		% within Age	86.3%	3.9%	3.9%	5.9%	100.0%

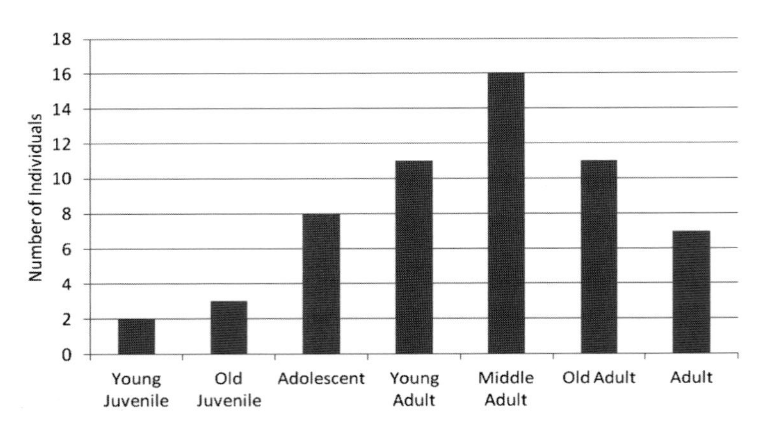

Figure 5.1. Isle of May age profile.

Biological sex could be determined for 52 skeletons. Methods used to determine sex include: visual assessment of morphological variation, specifically in the pelvis and skull (Bass, 1995) and metrical analysis (Bass, 1995). Stature was calculated using formulae developed by Trotter and Gleser (1952). Ortner (2003), Aufderheide and Rodríguez-Martín (1998), Waldron (2009), Mann and Hunt (2004) and Resnick (2002) were used in identification and diagnosis of skeletal lesions. Chi squared tests were run through the statistical package SPSS.

3. Results and Discussion

Since only 20% of the cemetery was excavated, the 58 skeletons are only a sample of the larger population. Although the burials are a good representation, interpretations offered here refer only to a portion of the entire population.

Demography of the Isle of May burials is seen in Figure 5.1. Of the 58 individuals, 52 could be sexed: 2 females, 1 probable female, 46 males, and 2 probable males. Six individuals are of unknown sex (of which half are un-sexed juveniles). The mostly male sex ratio is expected of this monastic community. However, the ratio could also indicate that females were buried elsewhere in the cemetery and remain unexcavated.

Table 5.4 and Figure 5.1 show the age profile for the Isle of May burials is as follows: 2 young juveniles, 3 old juveniles, 7 adolescents, 11 young adults, 16 middle adults, 11 old adults, and 7 who could only be broadly aged as adult. There are no infants or neonates among the articulated burials, but there are in the disarticulated

Table 5.5. Stature comparisons in Medieval Scottish and British lay and monastic sites.

Site Name	Avg. Male Stature (cm)	Avg. Female Stature (cm)	Reference
Ballumbie	168	157	This Study
Isle of May	170	-	This Study
Kinnoull Street	170	153	Bruce and Cox, 1995
Whithorn (Period V)	170	156	Cardy, 1997a
St. Mary's Church, Kirkhill	168	158	Bruce *et al.*, 1997
Logies Lane	169	157	Cardy, 1997b
Hallow Hill	169	160	Young, 1996
Aberdeen on the Green	170	158	Cardy, forthcoming
St. Nicholas Church (Phase A)	168	154	Duffy, forthcoming
Linlithgow	170	156	Cross and Bruce, 1989
Isle of Ensay	166	155	Miles, 1989
Glasgow Cathedral	172	156	King, 2002
St. Giles Cathedral (Period 2)	156	166	Henderson, 2006
Merton Priory	171	164	Waldron, 1985

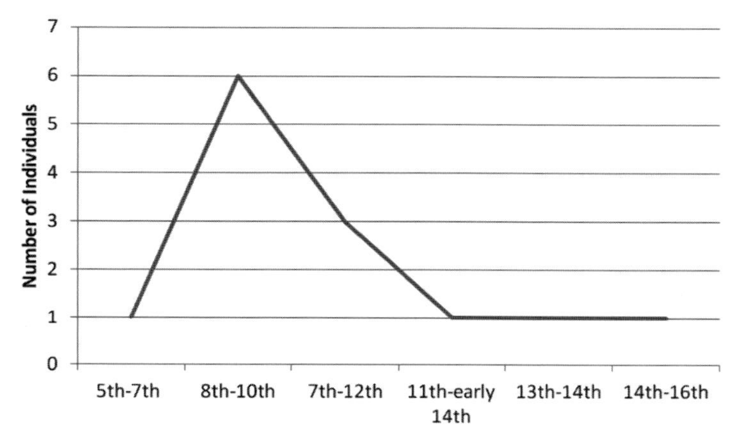

Figure 5.2. Isle of May 'Young Individual' burials per century. Note the overlap in centuries is intentional due to burial phasing by James and Yeoman (2008).

remains, suggestive of segregation in the cemetery and also perhaps a difference in burial practice.

If the Isle of May was solely a monastic community, mostly middle adult and old adult males would be expected. For example, Portmahomack in northern Scotland, a similar Pictish monastery, with a long cist burial ground dating to the 6th century, consisted of mostly older males (Carver, 2004). An English monastic site, Merton Priory, also had a low number of individuals less than 25 years of age (8%) (Waldron, 1985). The fact that 22% of the sample population at the Isle of May is under 25 years of age is surprising and reveals that the Isle of May was not only a monastic community, but may have served a function beyond normal monastic activities. The author suggests that the Isle of May was used as a place of healing, possibly a hospice for extreme cases, and that these younger adults visited the Isle of May specifically for that purpose. Seventy-five percent of these 'young individuals' were buried between the 8th and the 12th centuries, with a peak between the 8th and 10th centuries, which indicates that the height of healing tradition on the Isle of May was during that period (Figure 5.2). Although this healing tradition might have continued, the main function on the Isle of May shifts in the 12th century to a monastic one when the Benedictine priory is founded corresponding to the changing subadult percentages seen in Figure 5.2. While some monastic orders were known for medical treatment, the Benedictines did not practice medicine (Coulton, 1933), therefore the healing tradition on the Isle of May should not be associated with them but rather to earlier Celtic or even pagan occupations.

Of the sexed individuals, stature could be estimated for 46 skeletons; 45 males or probable males and 1 female or probable female. For males, stature ranged from 158.51 cm to 183.74 cm with a mean of 170.68 cm and a standard deviation of 5.9 cm. The stature for the only female where it could be estimated was 162.77 ± 3.66 cm. Since there is only a single stature for females, a mean could not be calculated. There are no male or female outliers for stature. Stature on the Isle of May is comparable with other Scottish and British medieval sites (Table 5.5).

Ninety-seven percent of the total Isle of May sample

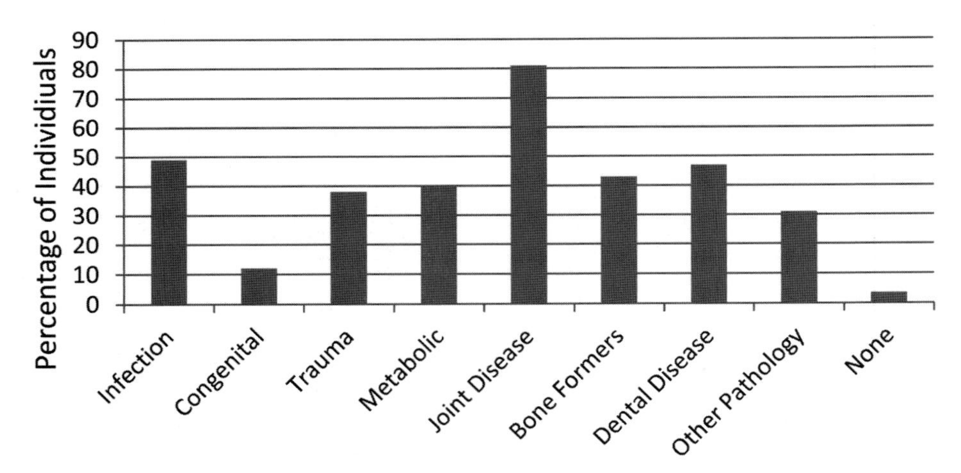

Figure 5.3. Pathology of total population at Isle of May.

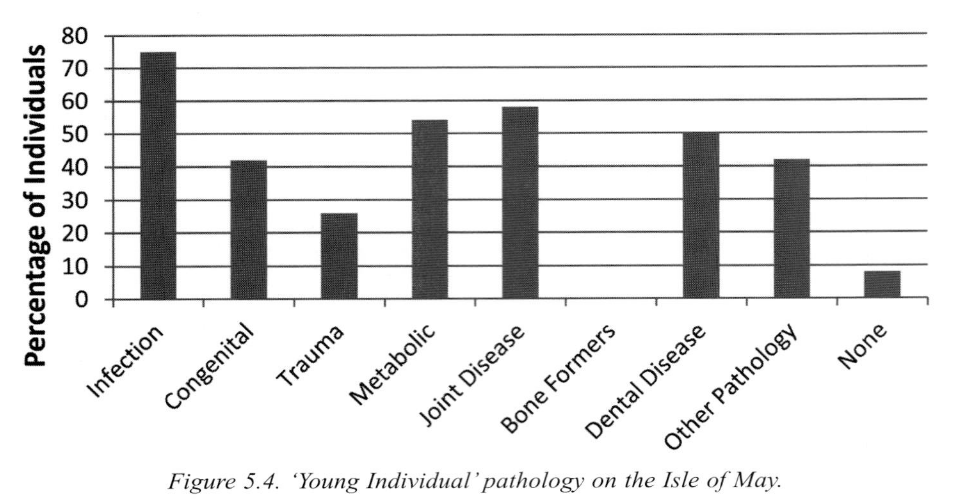

Figure 5.4. 'Young Individual' pathology on the Isle of May.

had at least one skeletal pathological lesion present. This extremely high prevalence clearly demonstrates that the Isle of May was not a normal population, even for one that was historically only monastic. For example in comparison, only 36% of the 676 individuals from the population at Merton Priory had pathology, a much lower statistic. Pathological conditions were placed into eight categories, which include: infectious disease, congenital disease, trauma, metabolic disease, joint disease, dental disease, and miscellaneous pathologies.

Of the total sample population at the Isle of May, 49% of individuals had evidence of infection, 12% had congenital disease, 38% had trauma, 40% had metabolic disease, 81% had joint disease, 43% are considered to be bone formers, 47% had dental disease, 31% had other pathology, and 3.5% had no skeletal pathology (Figure 5.3). These unusually high prevalence rates across all of the categories further suggest that individuals travelled to the Isle of May for healing purposes.

Although there were most likely older individuals who also travelled to the Isle of May for healing, it is interesting to focus on the individuals under 25 years of age, since it is more likely that the only purpose they had for travelling there was for healing, while older individuals could potentially also be either pilgrims or monks. Of the 'young individuals' sample, 75% had infection, 42% had congenital disease, 25% had trauma, 50% had metabolic disease, 58% had joint disease, none were bone formers, 50% had dental disease, 42% had other pathologies, and 8% had no skeletal pathology (Figure 5.4). These extremely high percentages of disease corroborate the theory that these 'young individuals' travelled to the Isle of May for healing and then were buried there when they succumbed to their disease.

3.1 Case Studies

To demonstrate the diseases found on the 'young individuals', a few case studies are included here. Figures 5.5–5.7 show the long bones of Skeleton 814, a 14–16 year old male. This individual had lesions consistent with Hypertrophic Pulmonary Osteoarthropathy (HPO), due to the large amounts of bilateral periostitis. Symmetrical periostitis is found on the femora, ulnae, and innominate bones. The

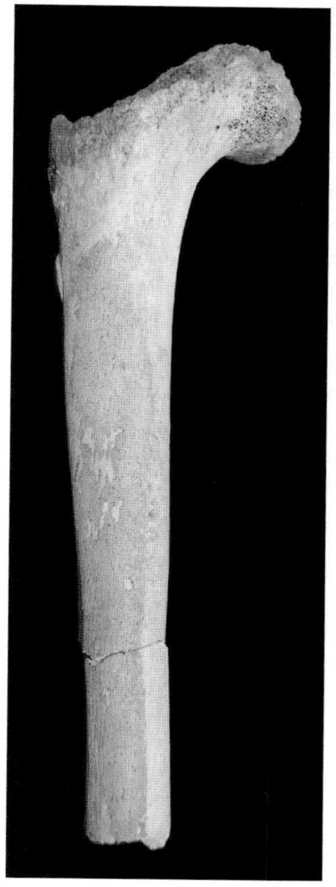

Figure 5.5. SK 814 right femur with diffuse periostitis, notice bilateral symmetry with left femur in Figure 5.6.

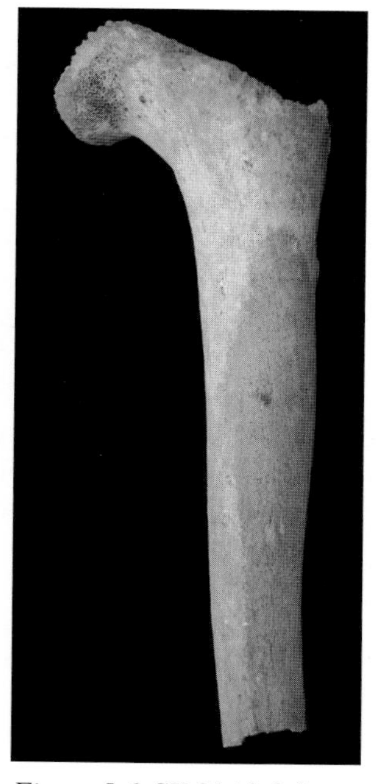

Figure 5.6. SK 814 left femur with diffuse periostitis, notice bilateral symmetry with right femur in Figure 5.5.

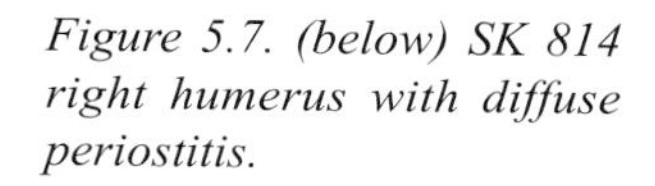

Figure 5.7. (below) SK 814 right humerus with diffuse periostitis.

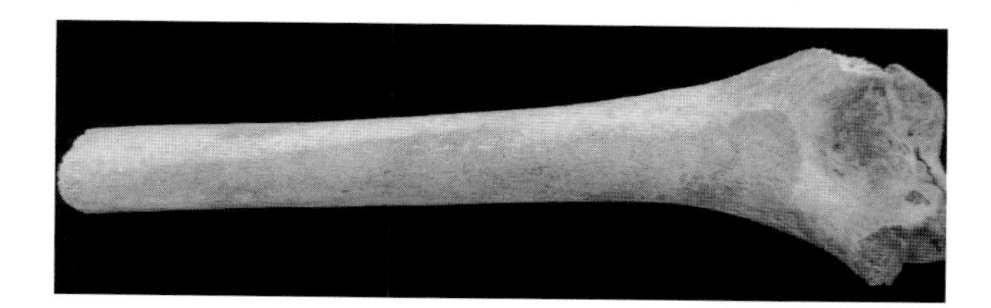

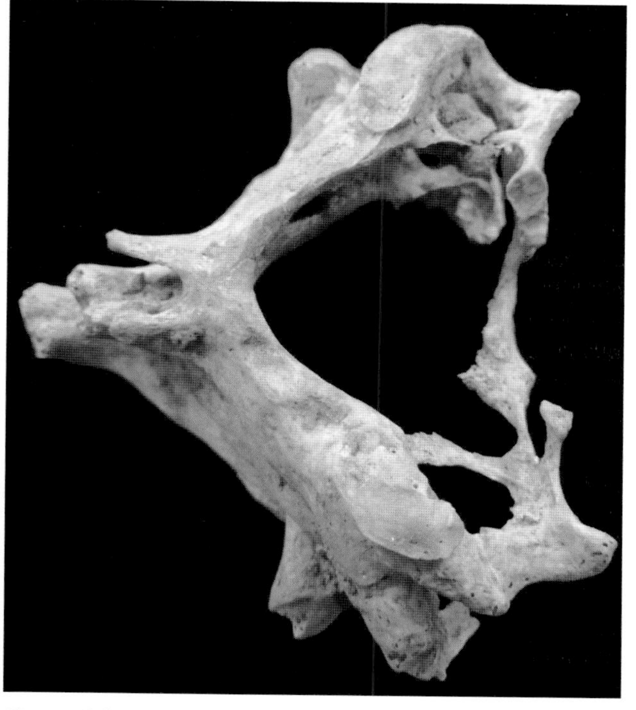

Figure 5.8. SK 972 tuberculosis of the spine; resorption of cervical bodies.

Figure 5.9. SK 972 tuberculosis of the spine; resorption of vertebral bodies and kyphosis.

right humerus has periostitis that wraps around most of the bone. Prevalence for HPO is often underestimated in archaeological burials due to the fragmentary nature of many assemblages (Mays and Taylor, 2002). HPO can be primary or secondary (Mays and Taylor, 2002; Resnick, 2002; Ortner, 2003), mostly affecting the long bone shafts, though in primary cases the periostitis tends to extend from the epiphyses (Rothschild and Rothschild, 1998; Resnick, 2002). Primary HPO is congenital in nature and generally affects prepubescent subadults, while secondary HPO affects older individuals suffering from chronic respiratory infections (Aufderheide and Rodríguez-Martín, 1998; Mays and Taylor, 2002). Tuberculosis is often the main cause of HPO in archaeological populations (Rothschild and Rothschild, 1998; Mays and Taylor, 2002; Ortner, 2003). Other causes are intrathoracic diseases and pulmonary infections due to air pollutants; in modern populations HPO is commonly associated with lung cancer and emphysema (Mays and Taylor, 2002; Resnick, 2002). Since Skeleton 814 is around puberty, it is difficult to diagnose whether he has primary or secondary HPO.

Dental age places Skeleton 972 at 8 years (± 2 yrs) while the length of the humerus and femur places this skeleton younger at 4.5–5.5 years old. It is likely that this child was 8 years old but was so sickly that the rest of the body was undeveloped. The pathology of the spine indicates tuberculosis; there is fusion and reabsorption of the bodies of many cervical and thoracic vertebrae, as well as a severe curvature of the spine (Figures 5.8 and 5.9). Tuberculosis is a common chronic infectious disease seen in many archaeological populations. It is caused by a species of mycobacterium (Aufderheide and Rodríguez-

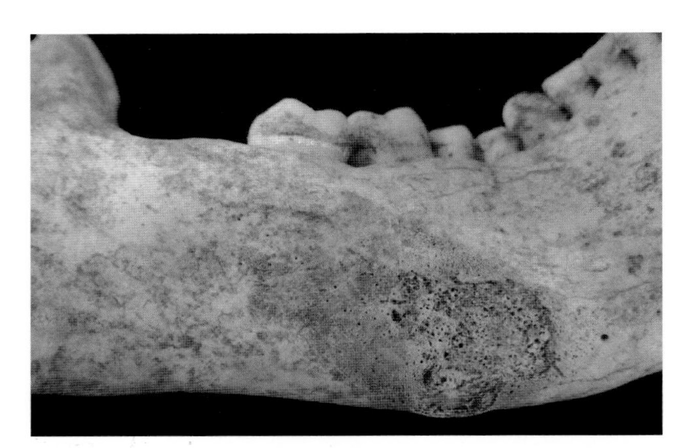

Figure 5.10. SK 997 periostitic lesion of the mandible.

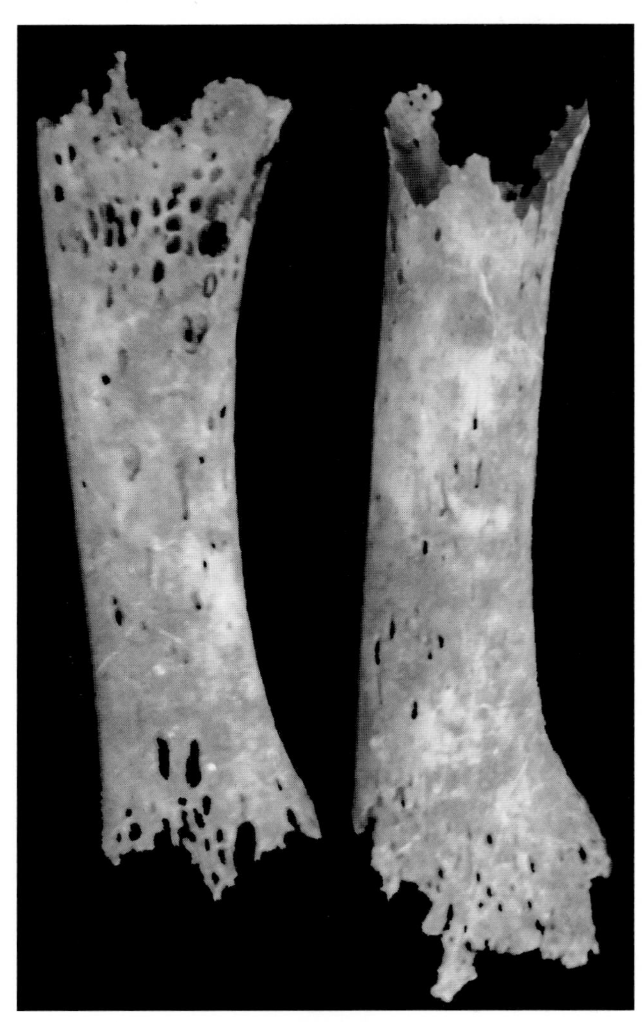

Figure 5.11. SK 997 dactylitis of the metatarsals.

Martín, 1998; Ortner, 2003). The most affected bones in individuals with tuberculosis are the vertebrae where destruction and resorption of bone often causes kyphosis, sometimes referred to as 'Pott's Disease' (Aufderheide and Rodríguez-Martín, 1998; Resnick, 2002). The spinal destruction of Skeleton 972 is so severe that the child would have been crippled. It is extremely unlikely that this child travelled to the Isle of May for any purpose other than for a chance to be cured of this debilitating disease, and due to the young age of the child, was most likely accompanied on the journey by a parent or other family member.

Skeleton 997 was a 15–17 year old probable male with a diagnosis that is still under debate. Differential diagnoses include tuberculosis and congenital syphilis. There are many active periostitic lesions throughout the body, but the legs bear the brunt of the disease, both femora are extremely thin from atrophy, both tibiae are swollen with osteomyelitis, although post mortem damage reduces visibility. The lower legs and feet contain no trabecular bone at all, rather are sheaths of cortical bone (Figure 5.11). Due to the young age of the individual and since the lesions would take many years to progress to this state, venereal syphilis was excluded from the possible diagnoses. If this individual did suffer from congenital syphilis, it may be one of the earliest cases in the UK. Recent radiocarbon information dates this skeleton to 640–710 CE. Evidence for pre-Columbian syphilis is discussed by Mays *et al.*, (2012); the individual is dated to 1050–1250 CE, possibly still positioning SK 997 as the earliest known case of congenital syphilis in the UK. Congenital syphilis is an infectious disease caused by Treponema pallidum passed down to the fetus by its mother in utero (Aufderheide and Rodríguez-Martín, 1998; Ortner, 2003). Tibiae are the bones most affected in congenital syphilis, with periostitis and osteomyelitis. Dactylitis of the finger and toes is also frequently observed in congenital syphilis with a 'formation of a thin and bony shell' (Aufderheide and Rodríguez-Martín, 1998: 296). Syphilitic dactylitis is often bilateral and symmetric (Resnick, 2002) and in Skeleton 997 both right and left metatarsals are affected. However, except for a lesion on the mandible (Figure 5.10), the skull is unaffected and the skull is often affected in cases of congenital syphilis (Aufderheide and Rodríguez-Martín, 1998; Ortner, 2003). The severity of this individual's disease would have been crippling, again indicating a parent or other family member accompanying them for healing purposes.

Each of these 'young individuals' had a fatal disease; it is very possible that they travelled to the Isle of May for some sort of health care, possibly with a parent, and when their sickness turned fatal, they were buried there.

3.2 The Isle of May in Context

To place the Isle of May sample in context it was compared with another Scottish medieval sample called Ballumbie (Figure 5.12). Ballumbie was a rural parish church in county Angus during the medieval period. It is now a residential area in the northeastern part of modern day Dundee. The cemetery was used for 1,000 years; the earliest radiocarbon date of the burials is 515 CE ± 35 years and the latest radiocarbon date is 1600 CE ± 35 years (SUAT report forthcoming). The excavation was done by Scottish Urban

Archaeology Trust in 2005, but remains unpublished. The skeletal remains are curated at the University of Edinburgh. Of the skeletal remains, there are 197 articulated burials, all of which were analysed by the author.

Ballumbie has a typical demographic profile with nearly a 1:1 male to female ratio (70 males, 77 females) (Figure 5.12). However, the Isle of May is a very atypical sample. There are only three females or probable females in the sample population, and few subadults which make the male to female ratio extremely high at 16:1. As mentioned earlier, the male biased sex ratio is expected from a monastic community and is comparable with other monastic sites such as Portmahomack and Merton Priory (Waldron, 1985; Gilchrist and Sloane, 2005; Carver, 2008).

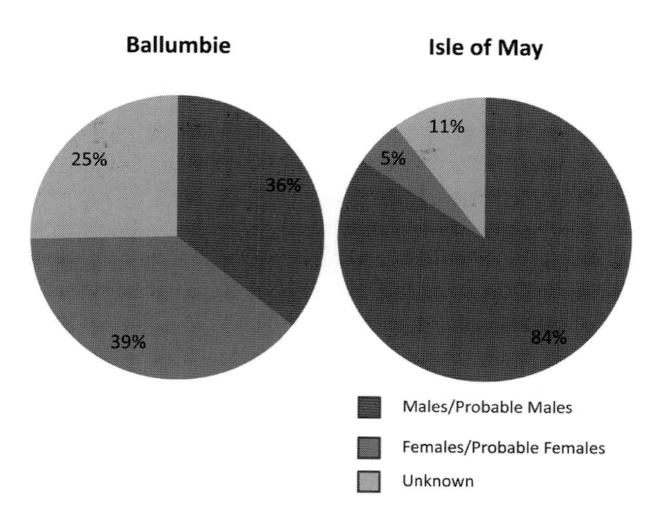

Figure 5.12. Isle of May and Ballumbie sex ratio comparison. Ballumbie has a nearly 1:1 male to female ratio where the Isle of May has a 16:1 male to female ratio.

The percentages of individuals in the different age categories for Ballumbie and the Isle of May are illustrated in Figure 5.13. Ballumbie has a very normal age profile with mostly middle adults but also the normal amount of subadults you would expect to find in a medieval cemetery (18% of the total population). Other medieval sites have comparable amounts of subadults; Horse Cross (20%), Hallow Hill (15%), Longniddry (13%), and Glasgow Cathedral (19%) (King, 2002; Lorimer, 1992; Young, 1996; Roberts, 2007). In medieval Scottish cemeteries subadult percentages generally range from 13% to 55%, seen at the Isle of Ensay (Miles, 1989). However, when the Isle of May is compared, it is clear that individuals there were older, with very few subadults (9%). When broadening the subadult range to include 18–24 year olds (young individuals) the percentage increases to 22% of the population. The individuals under the age of 25 are most likely the patients of the 'healing centre.' However, the large increase from 9% to 22% from juvenile to adolescence may suggest individuals were travelling to the Isle of May at that age (17–24 years) for both healing and monastic purposes or that more adolescents travelled there solely for healing purposes. The majority of the individuals in the total sample population at the Isle of May are middle adults between 35–45 years old and a large percentage of individuals (19%) lived into old adulthood. These older individuals most likely represent the monks. Individuals at Ballumbie also tended to die around middle adulthood; however fewer individuals survived into old adulthood (13%).

Ballumbie has a very typical prevalence for disease for a medieval sample with 60% of the sample having some form of pathology (Figure 5.14). On the other hand, the Isle of May has an extremely high prevalence for disease at 97% of the sample. When a chi squared test was run, it showed there is a significant relationship between the

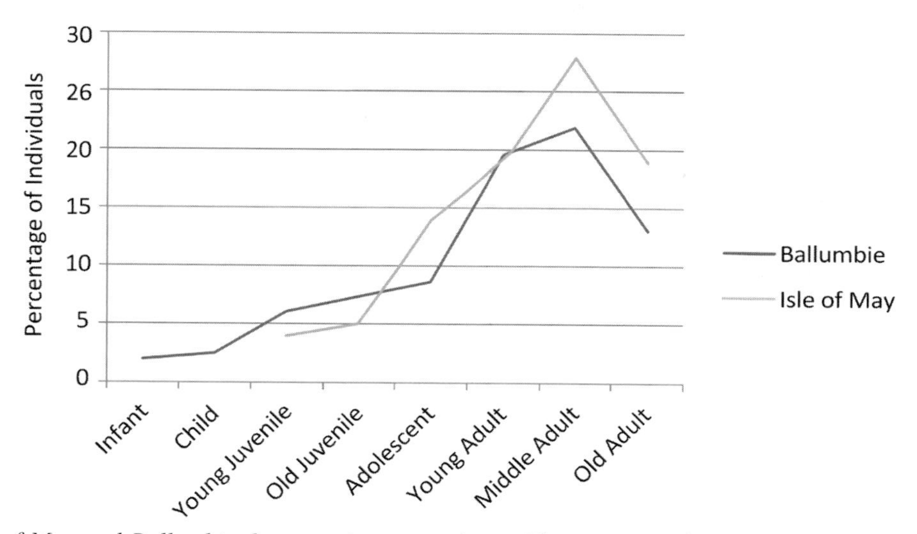

Figure 5.13. Isle of May and Ballumbie demography comparisons. The majority of individuals in both populations died in middle adulthood, however more individuals at the Isle of May lived into old adulthood.

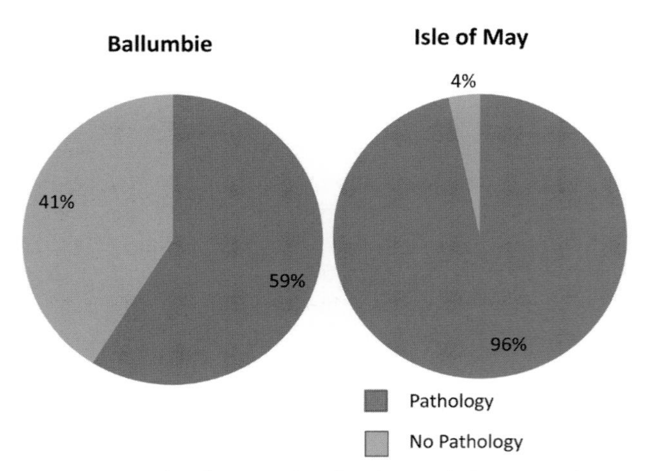

Figure 5.14. Isle of May and Ballumbie pathology prevalence. Just over half of the population at Ballumbie has pathology where at the Isle of May nearly the entire population has pathology.

pathology prevalence at Ballumbie and the Isle of May (χ^2=46.542, P<0.0001, df=1).

Figure 5.15 illustrates the percentages of each category of pathology for both samples. Ballumbie again shows a normal, healthy sample where the highest percentages are for joint and dental disease, both common in most archaeological populations (Aufderheide and Rodríguez-Martín, 1998; Ortner, 2003; Waldron, 2009). The sample at the Isle of May is abnormal since they have high percentages of almost every disease. Since more individuals lived into older adulthood than at Ballumbie (the monastic part of the community), it is expected to see a higher prevalence for joint disease at the Isle of May. However, the percentage of joint disease is high even when comparing with other monastic populations; Merton priory (67%) and Whithorn (56%) (Waldron, 1985; Cardy, 1997a). While the majority of individuals with joint disease at the Isle of May are adults, even juveniles and adolescents have joint disease as well, an unusual discovery. Joint disease is typically age related (Aufderheide and Rodríguez-Martín, 1998; Ortner, 2003; Waldron, 2009), therefore the joint disease in the younger individuals implies non-age related, but pathological causes. For example, one juvenile male at the Isle of May has Scheuermann's disease, a circulatory disorder that mostly affects male juveniles with wedge shaped vertebra and Schmorl's nodes.

Except for congenital disorders and dental disease, the Isle of May has drastically more of every other type of disease than Ballumbie. Dental disease is common in all medieval populations – generally age related and affecting all aspects of society, explaining the similar prevalence for dental disease in Ballumbie and the Isle of May (Aufderheide and Rodríguez-Martín, 1998; Ortner, 2003). The similar prevalence for congenital disorders at Ballumbie and the Isle of May suggests that individuals with these types of disorders did not travel to the Isle of May for healing; the amounts of congenital disorders found on the Isle of May reflect typical amounts found in other medieval populations (Cardy, forthcoming; Duffy, forthcoming; Bruce and Cox, 1995; Cardy, 1997b). Many congenital conditions, such as spina bifida occulta, are often asymptomatic and not terminal; therefore it is realistic that individuals would not make the journey for a cure for such a minor condition. Individuals that would be more likely to travel to the Isle of May for a cure would have serious pathologies such as infectious disease, metabolic disease, and fatal diseases that fall into the category of 'other disease'. This is the exact disease pattern that we see at the Isle of May, specifically in the 'young individuals' (Figure 5.4). While the conditions themselves might not necessarily be rare in other medieval populations, the severity of the disease most likely influenced the decision to travel to the Isle of May for a cure.

An interesting and surprising result is that all of those 'young individuals' at the Isle of May that could be sexed were male, which suggests only males travelled there for health purposes. This could signify preferential treatment in medieval Scottish society for males; drastic measures for sick males were taken such as travelling to religious sites for healing while perhaps females only received self-care or local folk healing. According to Gilchrist and Sloane (2005) differential healing practice is also seen at rural hospitals in England in the mid to late 12th century where there is a predominance of males, in some cases males make up 75% of the population. Other possibilities could be: females also travelled to the Isle of May but there was a differential burial practice; females that travelled to the Isle of May were not as severely sick as the males and survived.

Bone forming, or a propensity to ossify soft tissue, is also unrelated to a possible healing tradition. The aetiology of bone forming is still unclear, but recent studies suggest hormonal causes (Waldron, 2009; Rogers and Waldron, 2001). Most bone formers would be asymptomatic and never know they had the condition and thus would not have a need for a cure. Bone forming and its extreme condition, diffuse idiopathic skeletal hyperostosis, or DISH, is often associated with a monastic way of life, with high caloric intake as the culprit (Rogers and Waldron, 2001). While bone forming at the Isle of May could have been caused by a high calorie diet, there are only 2 cases that could be diagnosed as early onset diffuse idiopathic skeletal hyperostosis (eDISH). Another possibility is that individuals at the Isle of May could just have a predisposition to form bone.

Another way to look at prevalence for pathology is to count how many types of diseases each individual had during their life. For example, if a skeleton had infection and dental disease, their count is 2. The mean number of pathologies is significantly higher at the Isle of May (Table 5.6). This means the average individual buried on the Isle of May died with over 3 types of diseases while the average individual buried at Ballumbie died with just

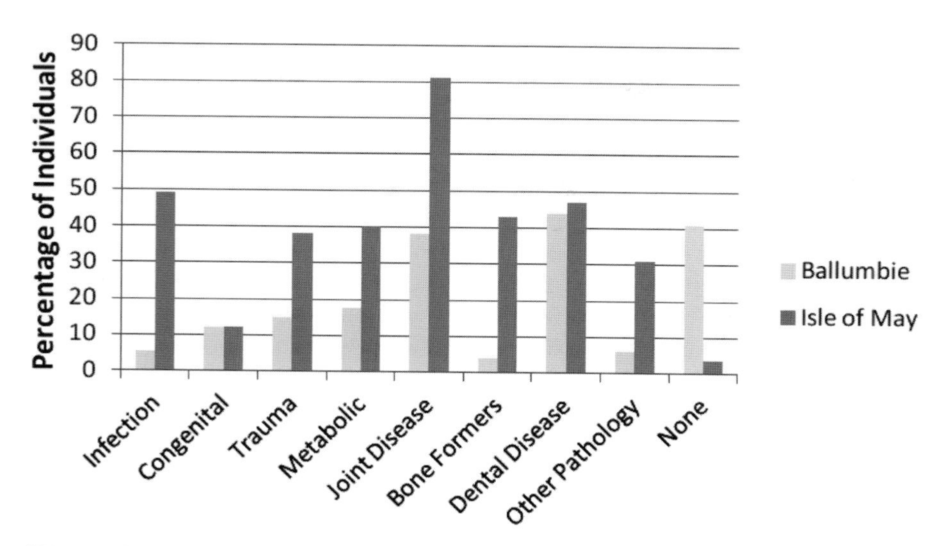

Figure 5.15. Isle of May and Ballumbie pathological condition comparisons. Except for congenital disorders, the Isle of May has higher amounts of all pathology categories.

Table 5.6. Number of pathological conditions per individual.

Population	Mean	Standard Error
Ballumbie	1.26	0.102
Isle of May	3.42	0.233

over 1 type. A chi-squared test revealed there is a significant relationship between the number of pathological conditions and population (χ^2=72.115, P<0.0001, df=7).

4. Conclusion

Woods *et al.* (1992) argue that the presence of pathological lesions on a skeleton actually signify the body's healthy response to disease. Conversely, the authors warn that a skeleton lacking any lesions may in fact signify an individual with a poor immune system who died before their skeleton could react to the pathogen (Woods *et al.*, 1992). It is, therefore, impossible to differentiate between an individual that died from an acute illness and a healthy individual who never suffered from disease during their lifetime.

However, when the Isle of May is compared to Ballumbie, many disease patterns emerge. The prevalence for disease is significantly higher in the Isle of May sample. There are higher percentages of nearly all types of pathology and, on average, individuals had more types of diseases during their lifetime. The drastically higher percentages of infectious disease, trauma, metabolic disease, joint disease, and other pathologies lead to the conclusion that some individuals buried on the Isle of May were clearly not part of the monastic community and travelled to the island for the purpose of healing. The medicinal plants henbane and greater celandine found in the environmental material on the Isle of May corroborate this conclusion. Black henbane has anesthetic properties and has been used medicinally for five millennia from ancient civilizations such as Egypt, Babylon, classical Greece and Rome. According to Moffat (1992) henbane was found at the medieval Scottish hospital at Soutra and was used in recipes to put a patient to sleep so "he will be able to suffer cutting in any part of the body without feeling it or aching." In an ointment henbane could be used for inflammation, burns, wounds, toothaches, earaches, and bruising (Moffat, 1992; Gruenwald *et al.*, 2000). Henbane was also infused into string and worn by children as a necklace to prevent convulsions and aid in teething (Moffat, 1992). Greater celandine was also used during the Middle Ages medicinally and was found in other monastic gardens (Harvey, 1992). It is a mild analgesic and was used to treat warts, inflammation, toothaches, gout, stomach, liver, gallbladder and intestinal problems (Gruenwald *et al.*, 2000).

The individuals that were buried on the Isle of May were obviously much sicker than the healthy population of Ballumbie. Due to the small size and precarious location of the Isle of May, it is likely that individuals travelling for a cure would have come from small, rural villages not too far from the Isle of May, perhaps villages such as Ballumbie. Health care in medieval Scotland consisted of travelling folk healers and surgeons in rural areas, and hospitals near larger burghs that mostly catered to lepers or the poor elderly (Hamilton, 2003). If an individual contracted a fatal disease outside the scope of the folk healers, they often travelled to locations with religious significance. It seems likely the Isle of May was one such location. The healing tradition at the Isle of May is connected to the island rather than the Benedictine priory, since the Benedictines did not practice medicine. The tradition instead predates the priory and is more likely tied to the earlier Celtic or pagan occupations at the site.

Perhaps the Isle of May had a reputation for curing specific diseases, and attracted individuals with those conditions. We do have evidence that barren women would travel to the Isle of May to drink the healing waters (Stuart, 1868). Another possibility is that the Isle of May catered to wealthy individuals who could afford to travel with a sick child to the island. The sick either travelled to the Isle of May, or were brought there by a parent because of its healing tradition or the burial ground was sacred and a preferable place to bury the dead. The historical accounts, the environmental evidence, and now the skeletal evidence all conclude that the Isle of May had a healing tradition during the medieval period.

Acknowledgements

I would like to thank the National Museum of Scotland, specifically David Caldwell and Jackie Moran for access of the Isle of May assemblage; Derek Hall and SUAT archaeology for their help with the historical and archaeological background to Ballumbie; Alison Cameron of Cameron Archaeology for unpublished reports; Isabella Kasselstrand for her help with SPSS; Kathleen McSweeney and James Fraser for their supervision; and Joe Nalven for his indispensable support.

References

Aufderheide, A.C., Rodríguez-Martín, C. 1998. *The Cambridge Encyclopedia of Human Paleopathology*. Cambridge University Press, Cambridge.

Bass, W. 1995. *Human Osteology: A Laboratory and Field Manual*, 4th edition. Missouri Archaeological Society, Columbia, MO.

Brooks, S., Suchey, J. 1990. Skeletal age determination based on the os pubis: A comparison of the Acsádi-Nemeskéri and Suchey-Brooks methods. *Human Evolution* 5, 227–238.

Brothwell, D. 1981. *Digging Up Bones: The Excavation, Treatment and Study of Human Skeletal Remains*, 3rd edition. British Museum (Natural History), London.

Bruce, M., Cox., A. 1995. The human skeletal remains from Kinnoull Street, in: Bowler, D., Cox, A., Smith, C. (Eds.), Four Excavations in Perth, 1979–84. *Proceedings of the Society of Antiquaries of Scotland* 125, 917–999.

Bruce, M., Cross, J.F., Kerr, N.W. 1997. The human remains, in: Rains, M., Hall, D., Bowler, D. (Eds.), *Excavations in St Andrews 1980–89: A Decade of Archaeology in a Historic Scottish Burgh*. Tayside and Fife Archaeological Committee Monograph 1, Glenrothes, pp. 118–131.

Byers, S. 2010. *Introduction to Forensic Anthropology*, 4th edition. Pearson, Boston, MA.

Cardy, A. 1997a. The environmental material. The human bones, in: Hill, P. (Ed.), *Whithorn and St Ninian: The Excavation of a Monastic Town, 1984–91*. Sutton, Stroud, pp. 51–62.

Cardy, A. 1997b. The human skeletal remains, in: Mackenzie, J., Moloney, C. Medieval Development and the Cemetery of the Church of the Holy Trinity, Logies Lane, St. Andrews. *Tayside and Fife Archaeological Journal* 3, 143–160.

Cardy, A. Human skeletal remains from Aberdeen on the Green, in: Cameron, A.S., forthcoming.

Carver, M. 2004. An Iona of the East: The early-medieval monastery at Portmahomack, Tarbat Ness. *Medieval Archaeology* 48, 1–30.

Carver, M. 2008. *Portmahomack: Monastery of the Picts*. Edinburgh University Press, Edinburgh.

Coulton, G.G. 1933. *Scottish Abbeys and Social Life*. Cambridge University Press, Cambridge.

Cross, J., Bruce, M. 1989. The skeletal remains, in: Stones, J.A. (Ed.), *Three Scottish Carmelite Friaries: Excavations at Aberdeen, Linlithgow and Perth, 1980–1986*. Society of Antiquaries of Scotland Monograph Series 6, Society of Antiquaries of Scotland, Edinburgh, pp. 119–142.

Daniell, C. 1997. *Death and Burial in Medieval England, 1066–1550*. Routledge, London.

Dilworth, M. 1995. *Scottish Monasteries in the Late Middle Ages*. Edinburgh University Press, Edinburgh.

Duffy, P. The human remains from the Kirk of St Nicholas Uniting, Aberdeen, in: Cameron, A.S., forthcoming.

Duncan, A.A.M. 1956. Documents relating to the Priory of the Isle of May, c. 1140–1313. *Proceedings of the Society of Antiquaries of Scotland* 90, 52–80.

Durkan, J. 1962. Cultural background in sixteenth century Scotland, in: MacRoberts, D. (Ed.), *Essays on the Scottish Reformation 1513–1625*. J.S. Burns and Sons, Glasgow, pp. 274–331.

Eggeling, W. 1960. *The Isle of May: A Scottish Nature Reserve*. Oliver and Boyd, Edinburgh.

Ewan, E. 1990. *Townlife in Fourteenth-Century Scotland*. Edinburgh University Press, Edinburgh.

Gilchrist, R., Sloane, B. 2005. *Requiem: The Medieval Monastic Cemetery in Britain*. Museum of London Archaeological Service, London.

Gruenwald, J., Brendler, T., Jaenicke, C. 2000. *PDR for Herbal Medicines*. Medical Economics, Montvale, NJ.

Hall, D. 2006. 'Unto yone hospitall at the tounis end': The Scottish medieval hospital. *Tayside and Fife Archaeological Journal* 12, 89–105.

Hamilton, D. 2003. *The Healers: A History of Medicine in Scotland*, 2nd ed. Canongate, Edinburgh.

Harvey, J.H. 1992. Westminster Abbey: The Infirmarer's Garden. *Garden History* 20, 97–115.

Henderson, D. 2006. The human bones, in: Collard, M., Lawson, J., Holmes, N. *Archaeological Excavations in St Giles' Cathedral Edinburgh, 1981–93*. Scottish Archaeological Internet Reports 22.

İşcan, M., Loth, S., Wright, R. 1984. Age estimation from the rib by phase analysis: White males. *Journal of Forensic Sciences* 29, 1094–1104.

James, H., Yeoman, P. 2008. *Excavations at St Ethernan's Monastery, Isle of May, Fife 1992–7*. Monograph 6. Tayside and Fife Archaeological Committee, Perth.

King, S. 2002. Human bone, in: Driscoll, S. *Excavations at Glasgow Cathedral, 1988–1997*. Society for Medieval Archaeology, London.

Lorimer, D. 1992. The human remains, in: Dalland, M. Long cist burials at Four Winds, Longniddry, East Lothian. *Proceedings of the Society of Antiquaries of Scotland* 122, 197–206.

Lovejoy, C.O., Meindl, R.S., Pryzbeck, T.R., Mensforth, R.P. 1985. Chronological metamorphosis of the auricular surface of the ilium: A new method for the determination of adult skeletal age at death. *American Journal of Physical Anthropology* 68, 15–28.

Mann, R.W., Hunt, D.R. 2004. *Photographic Regional Atlas of Bone Disease: A Guide to Pathologic and Normal Variation in the Human Skeleton*, 2nd edition. Charles C Thomas, Springfield, IL.

Mays, S., Taylor, G.M. 2002. Osteological and biomolecular study of two possible cases of hypertrophic osteoarthropathy from Mediaeval England. *Journal of Archaeological Science* 29, 1267–76.

Mays, S., Vincent, S., Meadows, J. 2012. A possible case of treponemal disease from England dating to the 11th–12th century AD. *International Journal of Osteoarchaeology* 22, 366–372.

Miles, A.E.W. 1989. *An Early Christian Chapel and Burial Ground on the Isle of Ensay, Outer Hebrides, Scotland with a Study of the skeletal Remains*. British Archaeological Reports, British Series 212, Oxford.

Moffat, B. 1992. SHARP Practice 4: *The Fourth Report on Researches into the Medieval Hospital at Soutra, Lothian Region*. SHARP, Edinburgh.

Ortner, D. 2003. *Identification of Pathological Conditions in Human Skeletal Remains*, 2nd edition. Academic Press, London.

Resnick, D. 2002. *Diagnosis of Bone and Joint Disorders*, 4th edition. W.B. Saunders, Philadelphia, PA.

Roberts, C.A., Cox, M. 2003. *Health and Disease in Britain: From Prehistory to the Present Day*. Sutton, Stroud.

Roberts, J. 2007. The human remains, in: Cox, A. Excavations at the Horse Cross, Perth. *Tayside and Fife Archaeological Journal* 13, 112–206.

Rogers, J., Waldron, T. 2001. DISH and the monastic way of life. *International Journal of Osteoarchaeology* 11, 357–365.

Rothschild, B.M., Rothschild, C. 1998. Recognition of hypertrophic osteoarthropathy in skeletal remains. *The Journal of Rheumatology* 25, 2221–2227.

Schaefer, M., Black, S.M., Scheuer, L. 2009. *Juvenile Osteology: A Laboratory and Field Manual*. Academic Press, London.

Stuart, J. 1868. *Records of the Priory of the Isle of May*. Society of Antiquaries of Scotland, Edinburgh.

Trotter, M., Gleser, G.C. 1952. Estimation of stature from long bones of American Whites and Negroes. *American Journal of Physical Anthropology* 10, 463–514.

Ubelaker, D. 1999. *Human Skeletal Remains: Excavation, Analysis, Interpretation*, 3rd edition. Taraxacum, Washington, DC.

Van Beek, G.C. 1983. *Dental Morphology: An Illustrated Guide*, 2nd edition. Wright, Bristol.

Waldron, T. 1985. *A Report on the Human Remains from Merton Priory*. English Heritage, London.

Waldron, T. 2009. *Palaeopathology*. Cambridge University Press, Cambridge.

Yeoman, P. 1999. *Pilgrimage in Medieval Scotland*. Batsford, London.

Young, A. 1996. The skeletal material, in: Proudfoot, E. Excavations at the Long Cist Cemetery on the Hallow Hill, St Andrews, Fife. *Proceedings of the Society of Antiquaries of Scotland* 126, 387–454.

6. First Evidence for Interpersonal Violence in Ukraine's Trypillian Farming Culture: Individual 3 from Verteba Cave, Bilche Zolote

Malcolm Lillie, Inna Potekhina, Alexey G. Nikitin and Mykhailo P. Sokhatsky

This paper presents the initial stages of an interdisciplinary study of human skeletal remains interred at Verteba Cave, western Ukraine. This site has been described previously as a "ritual site of the Trypillian culture complex" by Nikitin et al. *(2010), and the material considered here is one of seven crania recovered during excavations at Verteba between 2008 and 2010. Palaeopathological analysis of the individual considered here indicates that this is a young adult female with evidence for peri-mortem injury, cranial surgery and processing for interment. This evidence, together with the burial context itself, provides the first insights into early stage Trypillia culture inter-personal interactions and burial ritual in this region of Ukraine.*

Keywords: Violent interactions; Cranial surgery; Body processing

1. Introduction

This paper presents the results of recent fieldwork at Verteba Cave in Ukraine (Figure 6.1). This location is a unique site (as outlined below) containing human remains that date to the Eneolithic, and later, periods (Nikitin *et al.*, 2010). The site is located near the town of Bilche Zolote in southwestern Ukraine (Lat: 48.47/Long: 25.53), *ca.* 460 kilometres (290 miles) southwest of Kiev (Figure 6.1). The cave is an 8,555 metre long gypsum cave located in the Podillya region. Human remains have been excavated at this location since 1876, with more recent work undertaken by a team from the Borschiv Regional Museum, under the directorship of Mykhailo Sokhatsky and colleagues from Grand Valley State University. Absolute dating of disarticulated skeletal material (fourteen elements including a mandible, a number of ribs, vertebrae and a radius) from this cave are primarily dated to the Koshylovetska Trypillian (formally spelled "Trypolie" or "Tripolye") culture group (dated to phase CII, at *ca.* 3450–3100 BC) (Nikitin *et al.*, 2010: 11).

Whilst there are intrusive Linearbandkeramik (LBK) faming groups in the territory of Ukraine prior to the establishment of the Tripillian culture, this group represents the subsequent internal development of farming in Ukraine (Videiko, 1994; Zbenovich, 1996; Rassamakin 1999; Manzura 2005). Excavations in Romania have also identified the Cucuteni culture, which is named after the 'type-site' for this culture in this country (Zbenovich, 1996: 200). The Cucuteni-Trypillia culture developed through intensive contacts with, and expansion of, Neolithic Balkan-Carpathian agricultural groups (e.g. Körös, Boian, LBK, Hamangia and Petreşti) as they expanded northwards into Moldova and Ukraine (Korvin-Piotrovskiy, 2008: 24).

Nikitin *et al.* (2010) have previously noted that early stage burials/human remains are virtually non-existent until the CII phase of the Trypillia farming culture in the territory of Ukraine; i.e. after *ca.* 3400calBC (Dergatschov and Manzura, 1991). Indeed, Kruts (2008) reports that only individual finds have been made for the early and

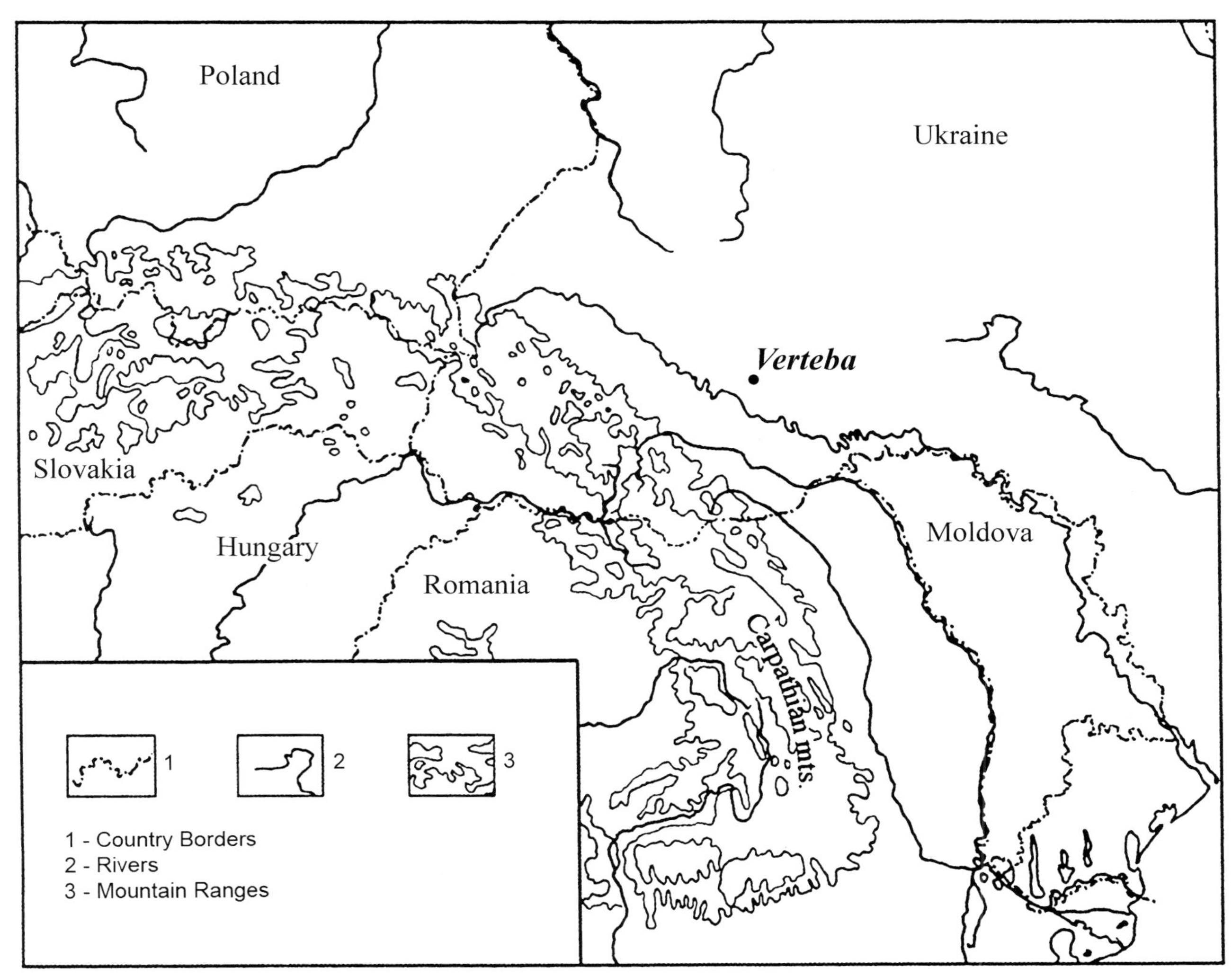

Figure 6.1. Map of the Podillya region with adjacent areas of Ukraine and neighbouring countries showing the approximate location of Verteba Cave (after Nikitin et al.*, 2010).*

middle stage of Trypillia, at sites such as Luka Ustynska, Soloncheny II, Vremye, Nezvysko and Lipkany, and that the majority of these finds are poorly preserved. As such, the dates that have been obtained on disarticulated skeletal elements from cultural horizons at Verteba are of particular significance, being earlier in some cases than finds of human skeletal material discovered for the Trypillia culture in Ukraine to date (Nikitin *et al.*, 2010: 9).

The cultural horizons themselves include significant numbers of pottery vessels (including conical dishes, deep bowls, bulging and bi-conical urns, and binocular-shaped vessels), anthropomorphic and zoomorphic figurines, faunal and human remains and charcoal (Nikitin *et al.*, 2010: 13). The matrix of the three cultural horizons in the cave includes loams with gravel, which were probably used to level the cave floor, carbonate clays mixed with charcoal and burnt gypsum, and some burnt wood and lime. Excavations since 1996 have shown that in excess of 1m of sediment has built up in the cave, with *ca.* 64 m^2 having been excavated to date, and the existence of sterile horizons between the occupation debris suggests that there are breaks in the use of the cave (*ibid.* 2010: 15). To date seven locations within the cave have been excavated, but it should be noted that the areas excavated are near the cave entrance, *ca.* 65m in from the entrance and *ca.* 95m in from the entrance, and that not all areas have continuous deposits in evidence.

2. Material and Methods

Analysis of the cranial remains at Verteba followed standard protocols for ageing and sexing as outlined in the literature (e.g. Ferembach *et al.*, 1980; Bass, 1987; Helmuth, 1988; Smith, 1984, 1991; Buikstra and Ubelaker, 1994). We recognise the limitations in ageing and sexing using cranial material, but as no reliable associations with postcranial material were identified during the excavations, or the current study, cranial morphology and dental wear were relied upon in this stage of the analysis.

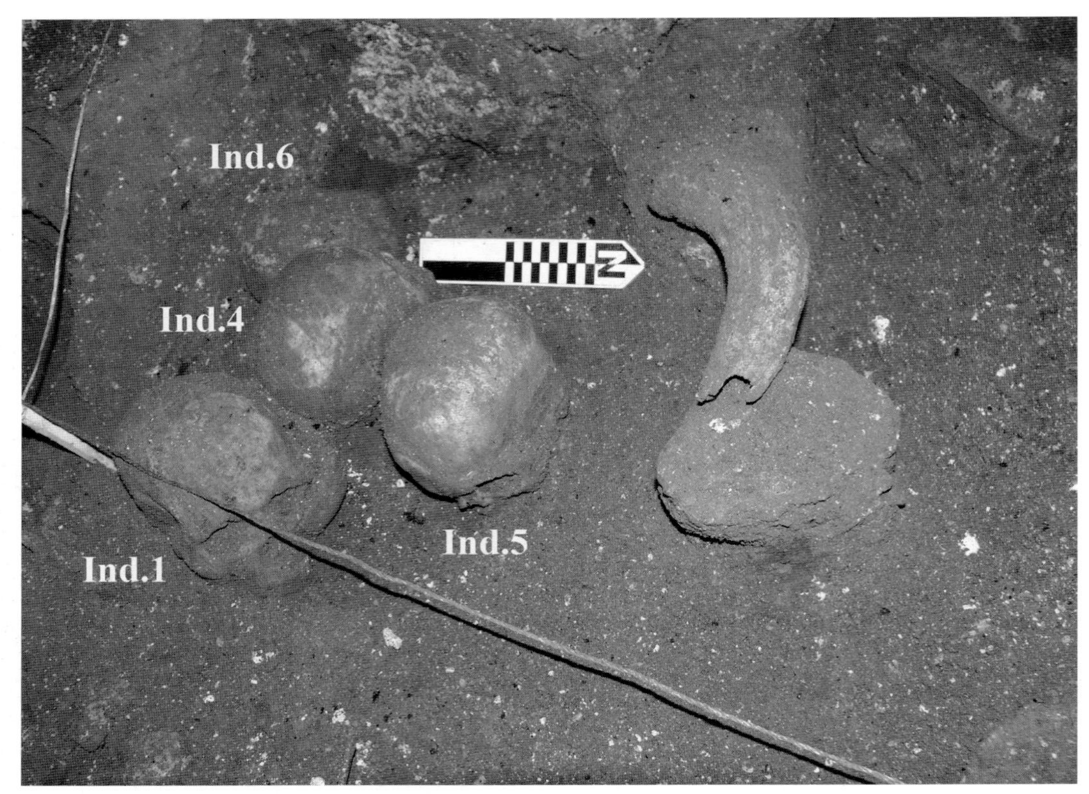

Figure 6.2. Crania Nest at Verteba Cave: four human crania to the left of the picture with aurochs horn resting on a large stone in the right-centre of image (image courtesy of M. Sokhatsky). Note: Individual 4 was shown to have the mandible in association, and this individual was also distinctive in that the humerus of a young child aged 6mths–1.5 years at death, was placed inside the crania prior to inclusion in the 'nest'.

All crania were located in close proximity to each other within the cave, with four of the crania located in a 'nest' (Figure 6.2) that was located in a niche, or recess, in the cave wall. The niche, which is located at the same level as the cave floor, was further demarcated from the general cave deposits by the construction of a low wall.

The remaining three crania were all located within *ca.* 1.4m of the 'nest' area, and given that access to this cave was unrestricted until recently, it is entirely possible that these crania could have formed part of the original deposit within the 'nest'. Individual 3 (discussed below) was recovered from the southeastern area of the earlier, 2007, excavations *ca.* 1.4m from the crania nest. The crania were all analysed by two of us (IP and MCL), and samples were recovered for AMS dating and stable isotope analysis of carbon and nitrogen at the Oxford facility, UK, with samples for strontium isotope analysis recovered for study by Professor Doug Price (Wisconsin). All of the crania have been measured, and recording of the nature of calculus deposition, caries expression (i.e. necrotic pits investigated macroscopically with a dental probe to assess severity of expression) and pathology has been undertaken. The results from the analysis of individual 3 are discussed below.

3. Results

Individual 3 (1.2SE) is a female? aged 18–22 (and probably towards the lower end of this range (Ferembach *et al.*, 1980; Bass, 1987). With the exception of damage in the right nasal area (including damage to the upper part of the maxilla, and right lacrimal, and the damage discussed below (and shown in Figure 6.5), the cranium is complete. In the maxillary dentition, the left first premolar, and first to third molars, and the right first premolar and first to third molars are all retained *in situ*, whilst the remainder of the dentition exhibits post-mortem loss (all sockets un-remodelled). Both first premolars and first molars show pinpoint dentine exposure, whilst the second and third molars are unworn, with only minor polishing on the second and third molars. The teeth have concretions of material adhering to them, making assessment of calculus deposition difficult to determine with any degree of certainty; this concreted material is probably derived from solution of the gypsum into the cave earth at this location as this precipitate is also visible on the crania from this cave. There is no indication of caries development on any of the preserved teeth of this individual.

The strontium and stable isotope data for diet requires processing, but a preliminary assessment of the data

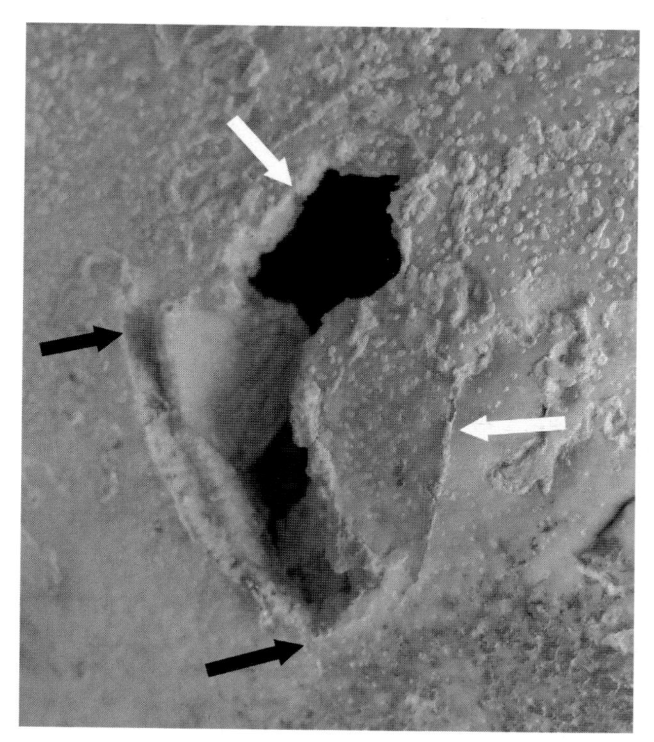

Figure 6.3. Impact damage on posterior of crania – Individual 3. The black arrows mark the sharp fracture line that is ca. 26mm in length, whilst the aperture is ca. 6mm wide at this point. The area of 'lost' vault is marked by the white top arrow which is ca. 15mm from the centre at the bottom left to the centre top right, this area appears to represent peripheral fracturing of the vault away from the main point of impact. The white right arrow marks the edge of the fracture on the right side of the aperture where the vault has been forced inwards by the pressure of the impact. Note calcite formation on vault and areas of damage (image © M. Lillie).

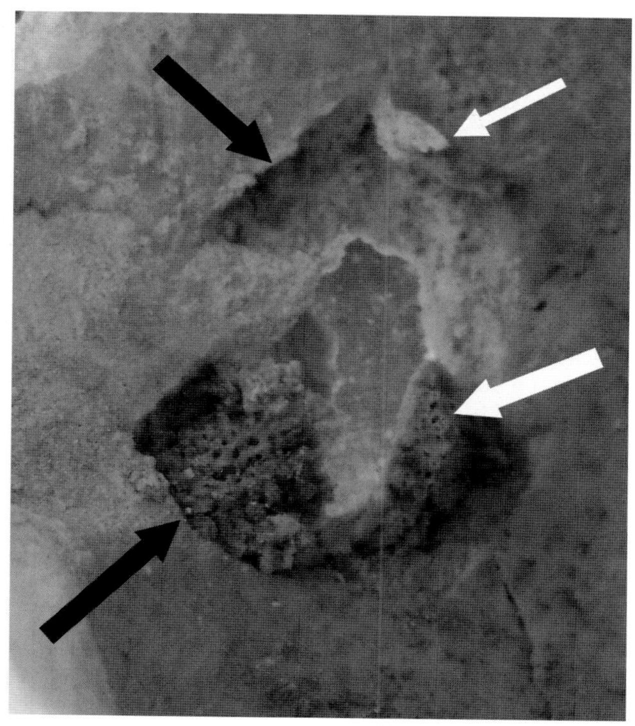

Figure 6.4. Internal view of impact damage on vault of Individual 3. White arrow shows fragments of bone that have been pushed inwards and remain in situ. Black arrows show areas where the inner surface of the vault has fragmented away due to impact pressure, and the top of the aperture has a small area of bone forced inwards (top white arrow), and peeling from the inner table of the vault (lower white arrow), indicating a fresh bone fracture (image © M. Lillie).

indicates that this individual is local to the region, as the strontium signature is the same as those obtained from pig molars and the other individuals from this cave. The dietary signal is consistent with farming, i.e. terrestrial C_3 diets, in which the associated fauna were not significant in dietary terms, i.e. the majority of the diet of these individuals was cereal-based. The data clearly indicates that this individual was part of a community that was engaged in agro-pastoral farming practices, as would be consistent with the Trypillian culture in this region. However, we should note that at present we are not in a position to assess any freshwater resource contribution to the diet. The AMS dating of this individual indicates an age of 3709–3537calBC (OxA-26202; 4863±33BP). This age is indicative of stage BII/CI of the Trypillia culture at 3900–3450calBC, making this one of the earliest dates yet obtained for Trypillia culture human remains from the territory of Ukraine.

Individual 3 has evidence for penetrating impact damage on the back of the vault at lambda. In Figure 6.3 it is apparent that the ectocranial surface (outer table) has been forced inwards by the impact, with blunt force compression in evidence and fragmentation of the vault (comminuted fracture); the ectocranial bone can clearly be seen still attached (*in situ*) on the left side of aperture.

In detail, the left side of the impact damage in Figure 6.3 is 26mm in length. At its widest point the entire insult is 32mm from left to right and *ca.* 38mm from top to bottom. It is difficult to interpret this damage without further study, but the position and structure of the damage would be consistent with a blow delivered from in front, and above of, a person who was kneeling forwards.

Figure 6.4 shows the internal detail of the trauma, with bone pushed inwards from the outer table of the vault in evidence alongside fracturing and 'peeling' of the endocranial surface, and loss of bone. The diploë is exposed, or lost, throughout the majority of the circumference of the aperture in the internal region, and the endocranial plate has fragmented away, leaving sharp, well defined edges around the aperture. The internal damage is more pronounced than the external damage (internally the maximum length is *ca.* 41.6mm and maximum width is *ca.* 33.2mm). There is no evidence for reactive bone in this area.

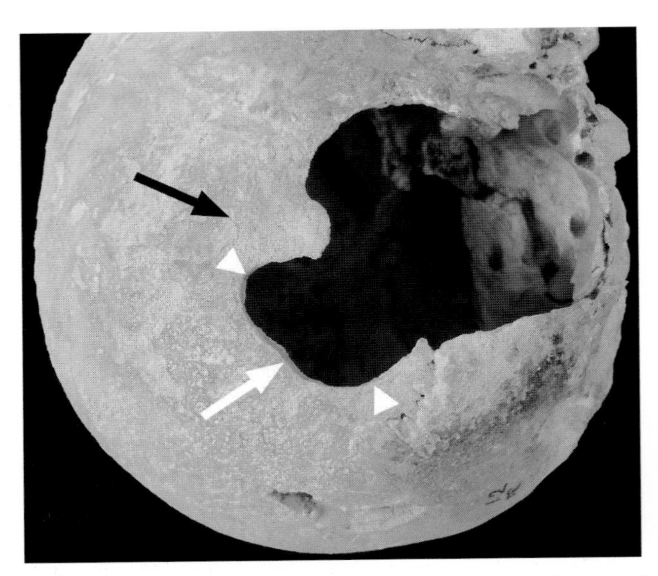

Figure 6.5. Individual 3 Probable trepanation – right side of crania showing sharp edges of aperture that indicate removal of bone (white arrow) and wear/erosion that has occurred in situ (black arrow). Distance between white triangles is 47mm (image © M. Lillie).

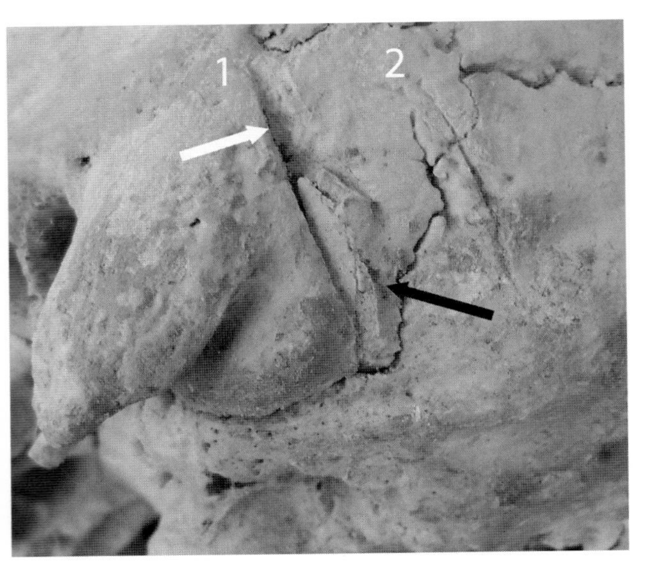

Figure 6.6. Cut marks on the left side of the cranium of individual No. 3 (1.2SE) from Verteba. Note fresh and sharp appearance of cuts located superior and posterior to the mastoid process indicating the cutting of the attachment of the neck muscles and tendons (in the area of attachment for Sternocleidomastoideus [SCM], Splenius capitis, and Longissimus capitis). The collapse and cracking of the bone (arrowed) indicates that processing occurred peri-mortem whilst the bone was still fresh. The laminar bone in the area of the white arrow has fragmented and fallen away from the vault whilst the bone in the area of the black arrow has remained in situ despite being compressed and fragmenting up from the vault (image © M. Lillie).

At this point it should be noted that there are no signs of bone reaction on this, or any of the other damage discussed in relation to individual 3, and similarly, all of the areas considered in this study were characterised as having exactly the same colour as the rest of the cranium. Furthermore, the calcite concretions in evidence over the surfaces of the vault also occur on all of the edges of the apertures, and other damage in evidence on this cranium.

The cranium of this individual also has evidence for the removal of bone (possible trepanation) from the right side of the vault (Figure 6.5); the fact that this bone removal was peri-mortem in nature is indicated by the very sharp edges to the aperture, visible 'saw' marks (although rodent gnawing cannot be discounted until further analysis is undertaken) along the surface of the cut, which are perpendicular to the cranial table throughout the area of the aperture that is preserved, and the fact that the cut edge is the same colour as the remainder of the vault. The only caveat to this observation is that the calcite deposition around and on the edge of the cut obscures any evidence for remodelling of the bone in this area.

The aperture appears to have comprised either two holes (in a figure of eight), or one figure of 8 shaped hole, and is irregular in shape. The edges of the cut are sharp and do not have any evidence for remodelling, perhaps reinforcing the suggested peri-mortem processing of the skull (with the above caveat in place). The absence of the temporal negates the recording of cut marks on this side of the vault (see below).

Cut marks are located on the left side of the vault, superior and posterior to the mastoid process, indicating cutting of the attachments of the neck muscles and tendons in order to detach the head from the body (Figure 6.6). The location of these cuts indicates the severing of the attachment of the neck muscles and tendons (in the area of attachment for Sternocleidomastoideus [SCM], Splenius capitis, and Longissimus capitis). Cut 1 is 27.84mm in length, and there is clear evidence for pressure damage towards the lower part of the incision, suggesting a left to right cutting action. Cut 2 indicates the use of less pressure and is 17.9mm in length, again with increased depth to the lower end of the incision.

As can be seen in Figure 6.6, the bone has clearly fragmented away from the vault (in the area of the white arrow) whilst in the area of the black arrow the pressure of the cut has caused the laminate to buckle upwards. The edges of these cuts are extremely clean and sharp, again suggesting peri-mortem activity. There appear to be suggestions of other cut marks on a number of the crania analysed at Verteba, but these initial observations will need to be confirmed in the laboratory.

4. Discussion

As the majority of the known burials of the Trypillian culture (over 20 sites to date) are associated with the

Kasperivska group (also known as Gorodsko-Kasperovsky or Gordyneshtsky type), which is dated to phase CII of Trypillia at *ca.* 3125–2775 BC (Nikitin *et al.*, 2010: 11), the dating of individual 3 to 3709–3537calBC, i.e. stage BII of Trypillia, is of particular significance. Similarly, the range of peri-mortem pathologies in evidence on individual 3 indicate that a combination of inter-personal violence, possible cranial surgery and the processing of this individual (removal of the skull) for inclusion in the burial context has been undertaken (e.g. Šlaus, 1994: 165; Orschiedt *et al.*, 2003; Pfeiffer and van der Merwe, 2004; Facchini *et al.*, 2008). The suggested removal of the crania for inclusion in the burial context would most likely have occurred soon after death, as it is clear from this study that the cut marks were made on fresh bone. The absence of the mandibles would appear anomalous given the removal of the skull as part of the interment ritual, and whilst no explanation is currently forthcoming from the current analysis this clearly warrants further investigation as the study of these deposits progresses.

The impact damage on the back of the crania of Individual 3 is pronounced, and clearly the result of an attack, perhaps aimed at causing grievous bodily harm. It is difficult to see how this injury could have occurred accidentally, although of course this scenario cannot be ruled out. The fact that this insult is peri-mortem in nature is reflected in the excellent preservation of the fractured vault pieces *in situ* and the total absence of evidence for remodelling of the vault. In addition to the pain and blood loss that the penetrating injury would entail, the neurological consequences, assuming that death was not immediate, would be considerable, and would have included brain damage as the ecto- and endocranial plates were forced inwards (cf. Orschiedt *et al.*, 2003: 380).

At present we have yet to determine whether the impact damage to the crania of Individual 3 was caused by a stone or metal implement. However, it should be noted at this point that whilst the Trypillian culture is an Eneolithic farming culture, these populations had access to copper metallurgy, so the weapon used to inflict the injury could be either stone or metal (probably an axe in either case). Equally, the knives(?) used in the processing of this individual (Figure 6.6) could have been of either type, and future study of the crania will include microscopic examination and XRF analysis of the cut surfaces to see if traces of copper can be detected, as and when funding allows.

In addition, to the above observations it is worth noting that this population had contacts, trade and exchange relations with contemporary central European groups such as the Tisza-Polgár and Baden culture groups, possibly due to the copper mining activities in the Verteba area (Nikitin *et al.*, 2010: 16; Sokhatskij *et al.*, 2000).

Finally, it is also apparent that the cutmarks on Individual 3 are distinct and lack evidence for re-modelling, indicating immediate post-mortem processing of the remains as part of the burial rituals undertaken by the BII phase Trypillia culture groups in the vicinity of Verteba Cave (Tung, 2008).

After discussions with colleagues during the BABAO conference in Bournemouth the lead author is inclined to consider the sequence of events as potentially comprising the initial violence to the person, causing the penetrating impact damage on the back of the vault. This would undoubtedly have been sufficient to warrant an attempt at intrusive surgery aimed at releasing the pressure that would have resulted from the injury, although initial removal of the impacted bone would have been a logical surgical approach. As such the possible evidence for cranial surgery on the right side of the vault may therefore be related to bone removal for ritual or trophy taking purposes, or both?

Subsequently, as with the other six cranial interred in this area of the cave, individual 3 was processed for inclusion in the burial context via removal of the cranium, as attested by the cut marks visible in Figure 6.6. The lack of evidence for healing of any of these three pathologies suggests that they were both peri-mortem and immediately post-mortem in nature.

5. Conclusions

The evidence for peri-mortem injury and surgery, post-mortem processing and the mortuary rituals undertaken on the human remains recorded at Verteba are unique for the Trypillia culture, both in terms of the dating, context, and range of activities in evidence. We are currently in the process of assessing the crania in relation to their anthropological type (craniometrics) and correspondence to the Trypillian culture composition (*cf.* Kruts, 2008), as the assignment of these crania to the Trypillian culture is currently based on the dating and context of the finds. Given the local strontium signatures and the burial context, this assumption is considered to be realistic at present. The full analysis of the anthropology and palaeopathology of these individuals, along with a consideration of their isotopic characteristics and burial context, is ongoing. However, it is already evident that this site is of considerable significance to our understanding of the early stages of the Trypillia culture farming population in Ukraine and to the elucidation of society in terms of inter-personal violence, possible cranial surgery or the removal of 'discs' as trophies, the processing of human remains in death, and the articulation of burial rites of these early farming groups.

Acknowledgements

We would like to thank the NERC-AHRC National Radiocarbon Facility (NRCF), Oxford for providing funding for the AMS dating (Project – No. NF/2011/2/18). We would also like to thank Sasha Potekhina for driving us from Kiev to Verteba (ML, IP), no small feat given the distance and condition of the roads in places! Colleagues at Bilche Zolote are also thanked for their help with the fieldwork. Alexey Nikitin and Gwyn Madden, and students from Grand Valley State University, have undertaken excavations at Verteba since 2008, and the material discussed in this paper comes from a visit undertaken by two of us (IP, ML) in 2011. Funding for the fieldwork was provided by the Wetland Archaeology and Environments Research Centre at the University of Hull. ML would like to thank Chris Knüsel for earlier comments on the pathologies discussed and Chris Meiklejohn (Winnipeg) and Deborah Merrett (SFU) for additional thoughts on this material. As always any errors or omissions rest with the lead author.

Finally, we would like to dedicate this paper to the memory of a number of individuals, who were friends, colleagues and mentors, including; Professor Dimitri Telegin, Professor Marek Zvelebil, Professor Vladimir Timofeev, Professor Pavel Dolukhanov, Dr Ken Jacobs and Dr Roger Jacobi.

References

Bass, W.M. 1987. *Human Osteology: A Laboratory and Field Manual*. Archaeology Society, Columbia, MO.

Buikstra, J.E., Ubelaker, D.H. 1994. *Standards for Data Collection from Human Remains*. Arkansas Archeological Survey Research Series No.44, Fayetteville.

Dergatschov, V.A, Manzura, I.V. 1991. *Pogrebal'nye kompleksy pozdnego Trypolya (Burial Complexes of the Late Trypillia)*. Shytnytsa, Kishinew.

Facchini, F., Rastelli, E., Belcastro, M.G. 2008. *Peri mortem* cranial injuries from a Medieval grave in Saint Peter's Cathedral, Bologna, Italy. *International Journal of Osteoarchaeology* 18, 421–430.

Ferembach, D., Schwidetzky, I., Stloukal, M. 1980. Recommendations for age and sex diagnosis of skeletons. *Journal of Human Evolution* 9, 517–549.

Helmuth, H. 1988. *A Laboratory Manual in Physical Anthropology*. Canadian Scholar's Press, Toronto.

Korvin-Piotrovskiy, O.H. 2008. Trypilian culture in Ukraine, in: Ciuk, K. (Ed.), *Mysteries of Ancient Ukraine: the Remarkable Tripilian Culture 5400–2700 BC*. Royal Ontario Museum, Canada, pp. 22–31.

Kruts, S.I. 2008. Anthropological composition of the Trypillian population, in: Ciuk, K. (Ed.), *Mysteries of Ancient Ukraine: the Remarkable Tripilian Culture 5400–2700 BC*. Royal Ontario Museum, Canada, pp. 55–57.

Manzura, I. 2005. Steps to the steppe: Or, how the North Pontic region was colonised. *Oxford Journal of Archaeology* 24, 313–338.

Nikitin, A.G., Sokhatsky, P., Kovaliukh, M.M., Videiko, M.Y. 2010. Comprehensive site chronology and ancient mitochondrial DNA analysis from Verteba Cave – a Trypillian culture site of Eneolithic Ukraine. *Interdisciplinaria Archaeologica: Natural Sciences in Archaeology* 1, 9–18.

Orschiedt, J., Häußer, A., Haidle, M.N., Alt, K.W., Buitrago-Téllez, C.H. 2003. Survival of a multiple skull trauma: The case of an Early Neolithic individual from the LBK enclosure at Herxheim (Southwest Germany). *International Journal of Osteoarchaeology* 13, 375–383.

Pfeiffer, S., van der Merwe, N.J. 2004. Cranial injuries to Later Stone Age children from the Modder River Mouth, Western Cape Province, South Africa. *South African Archaeological Bulletin* 59, 59–65.

Rassamakin, Y. 1999. The Eneolithic of the Black Sea steppe: Dynamics of cultural and economic development 4500–2300 BC, in: Levine, M., Rassamakin, Y., Kislenko, A., Tatarintseva, N. (Eds.), *Late Prehistoric Exploitation of the Eurasian Steppe*. McDonald Institute Monographs, Cambridge, pp. 59–182.

Šlaus, M. 1994. Osteological evidence for peri-mortem trauma and occupational stress in two Medieval skeletons from Croatia. *Collegium Anthropologicum* 18, 165–75.

Smith, B.H. 1984. Patterns of molar wear in hunter-gatherers and agriculturalists. *American Journal of Physical Anthropology* 63, 39–56.

Smith, B.H. 1991. Standards of human tooth formation and dental age assessment, in: Kelley, M.A., Larsen, C.S. (Eds.), *Advances in Dental Anthropology*. Wiley Liss, New York, pp. 143–68.

Sokhatskij, M., Klochko, V., Manichev, V., Kvasnista ,V., Kozak, S. and L. Demchenko. 2000. Issues concerning Tripolye metallurgy and the virgin copper of Volhynia. *Baltic-Pontic Studies* 9, 168–186.

Tung, T.A. 2008. Dismembering bodies for display: A bioarchaeological study of trophy heads from the Wari site of Conchopata, Peru. *American Journal of Physical Anthropology* 136, 294–308.

Videiko, M.Y. 1994. Tripolye – "Pastoral" contacts, facts and character of the interactions: 4800–3200 BC. *Baltic-Pontic Studies* 2, 5–28.

Zbenovich, V.G. 1996. The Tripolye culture: Centenary of research. *Journal of World Prehistory* 10, 199–241.

7. Beheading at the Dawn of the Modern Age: The Execution of Noblemen during Austro-Ottoman Battles for Belgrade in the Late 17th Century

Nataša Miladinović-Radmilović and Vesna Bikić

The early modern history of Belgrade, Serbia, was marked by battles between the Ottoman and Habsburg empires, but also by extensive reshaping of the fortress and the town. The changes that Belgrade underwent are attested by rescue excavations on the Rajićeva Street site outside the fortress walls. Trench 12 was particularly interesting to archaeologists and anthropologists. A single location yielded five bodiless skulls. All belonged to males, aged between 20 and 45 years of age. All showed decapitation marks, apparently caused by a sword cut at the level of cervical vertebrae (C2, C3 and C4), except one, inflicted at the lower third of the occipital bone and severing part of the right mastoid process. There was no evidence that the heads had been publicly displayed, except in one case, where both the skull base and C1 had suffered additional violent injuries when impaled. The skulls had probably been carefully disposed of after the execution, as evidenced by the presence of mandibles and anatomic connection of cervical vertebrae which had not yet succumbed to postmortem decomposition. From all available data, the interrelationship of 'burials' and structural remains, and coin finds, the beheading may be placed between 1688 and 1717.

Keywords: Perimortem injuries; Sharp blade cut marks; Decapitation

1. Introduction

Beheading was widely considered the severest penalty imposed only for the most serious crimes, such as an act of treason against a state or a sovereign. Unfailingly ending in death, it is a method of execution, not of torture. In some societies, such as England, beheading with a sword or an axe was considered an honourable death and was reserved for the nobility, whereas commoners and the poor were more often sentenced to death by hanging (Daniell, 1997). At the dawn of the modern age this form of capital punishment was in use in the Ottoman Empire as well (Wiltschke-Schrotta and Stadler, 2005: 58–59). Sporadic information has come down to us from European diplomatic travellers visiting the Balkans in the 16th and 17th centuries. Some of them claimed to have seen severed heads of spies put on public display as a deterrent to others (Levental, 1989: 61). The most gruesome of such accounts is certainly that of the execution of Nicolas Doxat de Morez, the Austrian army colonel and military engineer who served as head of the Construction Department in Austrian-held Belgrade in 1723–36 (Leben des Herren Baron Doxat von Morez, 1757: viii–xii; Popović, 2006: 219–242). For the abortive defence and surrender of the fortress of Niš to the Ottomans, he was sentenced to death by beheading and executed in Belgrade, not far from the fortress: 'The headsman, who showed up in the meantime, struck an unfortunate blow, which cut too deep into his shoulder, and he tumbled off the chair without letting out a slightest scream. On the ground, it was only the fourth blow that cut off his head' (Leben des Herren Baron Doxat von Morez, 1757: 65–67).

Although material evidence for such executions is seldom found, there are examples from virtually all periods of the past (Harman *et al.*, 1981; Wells, 1982; Anderson, 2001; Walker, 2001: 588–590; Wiltschke-Schrotta and Stadler, 2005). To the best of our knowledge, one such skull was archaeologically recovered from the area just outside the southeast fortress wall, but it has not received due attention because of an unclear find context (a levelling layer).[1] Yet, the find-spot seems to suggest that the head was put on display, possibly at the top of the fortress wall. The five skulls presented here are the only archaeological discovery from this period in the Balkans to date attesting to the death penalty by beheading. In addition to blade cut marks on the bones, which permit reconstruction of the beheading technique, its significance resides in a clear and well-defined context.

1.1 Context of the Discovery

The location of Rajićeva Street site is significant in itself (Figure 7.1a), because it is an area outside the line of fortifications which had a somewhat different development from the fortress through the stages of Belgrade's urban transformation from the time of Roman dominance (2nd–4th c.) until the 19th century. In view of the stratigraphy of the site, we focused on the last Austro-Ottoman horizon. It began with the Ottoman conquest of Belgrade in 1521 and lasted practically until the final Ottoman withdrawal in 1867. Especially dynamic was the period between the 1690s and the late 18th century, marked by Austro-Ottoman wars and extensive reshaping of the fortress and the town. The extent and nature of these changes is clear from the surviving siege and rebuilding plans (Škalamera, 1973a, 1973b, 1973c, 1975a, 1975b, 1975c, 1975d, 1975e, 1975f), and largely attested by archaeological excavations. In that sense, the archaeological context of the discovery is substantiated by those archival sources.

The rescue excavations conducted in 2004 uncovered the skulls (G 20–24, i.e. Nos. 1–5) and nine skeletons in the Austro-Ottoman layer, which, in that part of the site, overlies a late Roman street that led from the Roman castrum to the civilian settlement. The skulls were grouped together and, as it appeared at the moment of discovery, three were laid on the left cheek and two on the right, with two of the five partially overlying each other (Figure 7.1b). In their immediate vicinity were nine skeletons, six male and three female, of different ages. Apparently, the bodies had simply been laid on the ground (no burial pits were identified) and variously orientated, though mostly northeast to southeast and west to east; in addition, four male skeletons (G 13–16) were strangely intertwined. This cluster of osteological remains was dated by the coins found in the layer, the youngest of which was a Ragusan coin of 1684.

The five skulls and nine skeletons were disturbed and partly damaged by the foundation wall of a later building, which, judging from the surviving plans of the fortress and town of Belgrade, could have been either of the following two barracks: Austrian Builders' (*Maurer*) Barrack, built in 1727 to accommodate the engineer units engaged in construction work on the fortress and around the town (Škalamera, 1975c: 23–25), or Ottoman Sipahi Cavalry Barrack, built on the site of the former, which had been razed to the ground after the Ottoman takeover of Belgrade in the autumn of 1739 (Škalamera, 1975d). Judging by the construction method and size of the building, we are inclined to identify it as the Ottoman Sipahi Barrack.

The approximate date and circumstances of the 'burial', however, can be assumed from the analysis of historical plans with high certainty. Of relevance to establishing the terminus post quem are the plans from the period of Austro-Ottoman wars (1688–1690) and the ensuing Ottoman reshaping of the town (1690–1717). The oldest plan (1688) shows on this site a musalla – an enclosed open air area for prayer – and, around it, a large cemetery (*Orta mezarlik*) for Muslim soldiers killed in the battle for Belgrade in 1521 (Škalamera, 1975a: 18–21). Excavations in the east zone of the site unearthed a part of this cemetery with individual and communal burials, and fragments of three Islamic grave markers. However, the cluster of skulls within the former musalla enclosure indicates that they could not have been buried there until after the prayer space was abandoned or demolished, which may be related to the Austrian capture of Belgrade in 1688 (Škalamera, 1975b: 22). On the other hand, the terminus ante quem would be the Austrian reshaping of Belgrade that began in 1717. From all available data, the interrelationship of osteological and structural remains, and coin finds, the date of the beheading may be limited to a period between 1688 and 1717.[2]

2. Materials and Methods

This paper will be confined to the anthropological analysis of the five skulls (Nos. 1–5, Table 7.1). In determining the sex of the beheaded individuals, we focused on morphological features of the cranium (*glabella*, *planum nuchale*, *processus mastoideus*, *processus zygomaticus*, *arcus supercilialis*, *protuberantia occipitalis externa*, *os zygomaticum*, *tubera frontale et parietale*, *os frontale* slope angle, *margo supraorbitalis* and shape of *orbitae*) and mandible (general appearance: *corpus mandibulae*, *ramus mandibulae* and *angulus mandibulae*; *mentum*, *angulus mandibule* and *margo inferior*), using the method established by a group of European anthropologists (Ferembach *et al.*, 1980: 523–524) and Buikstra and Ubelaker (1994: 19–21). Age estimation tools used were the changes on the occlusal surfaces of all teeth according to Lovejoy's scoring of age-related occlusal wear (Lovejoy, 1985).[3]

All necessary measurements for calculating cranial and mandibular indices were made (Bass, 1995).[4] Cranial and mandibular metrical elements and indices are presented in Table 7.2 for each skull. On teeth, mesiodistal and buccolingual diameters were measured and they are presented in Table 7.3 as recommended by Hillson (1990: 240–242, 1996: 80–82). Dental analyses (Tables 7.4–7.6), palaeopathological analyses and observation of non-metric variations (Table 7.7) were also conducted.

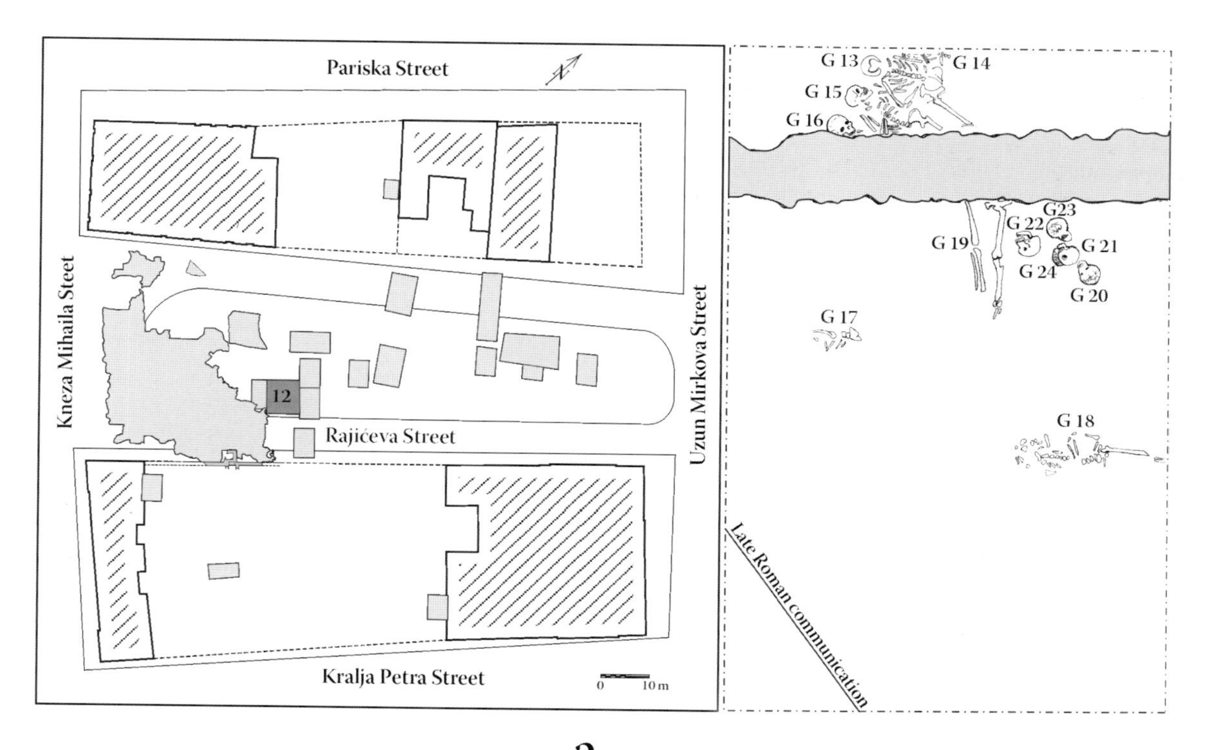

Figure 7.1. a) Location of Rajićeva street site, with find-spot of skulls (drawing: N. Lazarević). b) Skulls during excavation (photo: S. Pop-Lazić).

Table 7.1. Inventory of preserved bones.

PRESERVED BONES OF CRANIAL SKELETON	G 20	G 21	G 22	G 23	G 24
Frontal bone	75–100%	75–100%	75–100%	75–100%	50%
Right parietal bone	75–100%	75–100%	75–100%	75–100%	75–100%
Left parietal bone	75–100%	75–100%	75–100%	75–100%	75–100%
Right temporal bone	75–100%	almost 100%	75–100%	75–100%	75–100%
Left temporal bone	75–100%	100%	75–100%	100%	50–75%
Occipital bone	75–100%	100%	75–100%	75–100%	75–100%
Right mastoid process	100%	100%	75–100%	100%	100%
Left mastoid process	100%	100%	100%	100%	100%
Right zygomatic bone	100%	100%	-	100%	-
Left zygomatic bone	100%	100%	75–100%	100%	50%
Hyoid bone	50–75%	-	-	-	-
Skull bone fragments	-	-	6	15	27
Fragments of skull base bones	31	36	42	19	51
Maxilla	75%	75–100%	50%	75%	75–100%
Mandible	75–100%	75–100%	75–100%	75–100%	75–100%
POSTCRANIAL SKELETON					
C1	3 fragments	100%	fragment	fragment	100%
C2	dens missing	50–75%	-	100%	75–100%
C3	almost 100%	-	-	100%	-
C4	-	-	-	75%	-

3. Results

Skull No. 1 (G 20)

Skull No. 1 belonged to a male aged 30–45 (Tables 7.1–7.7).

Sharp force traumata consistent with decapitation are observable on C3 (Figures 7.2a and 7.2c) and the right gonion (Figures 7.2b and 7.2c). The cut on C3 only nicked the *processus articularis inferior* on the right side, while affecting on the left side *processus articularis inferior*, *arcus vertebrae* (*processus spinosus* is completely missing) and *corpus vertebrae* (at an angle of 30° to the vertical of the spinal cord).

Another observable palaeopathological change is an osteoma of 0.6 cm diameter in the middle of the frontal bone above the glabellar region.

Skull No. 2 (G 21)

Skull No. 2 belonged to a male aged 20–24 (Tables 7.1–7.7).

The left parietal bone shows two incidences of sharp force trauma: one that almost bisected the bone (downward blow at an angle of 80°, Figure 7.3b), the other parallel to the lambdoid suture (angle of 45°) (Figure 7.3a).

Sharp force trauma consistent with decapitation is observable on the left gonion (downward blow at an angle of 65°, Figure 7.3c) and C2 (Figure 7.3d). The body of C2 is bisected at an angle of 43° to the vertical of the spinal cord (Figure 7.3e).

Other palaeopathological changes are an injury with associated infection in the *spina nasalis* area, and an anomaly in the development of the atlas (*foramen arteriae vertebralis* instead of *sulcus*).

Skull No. 3 (G 22)

Skull No. 3 belonged to a male aged 25–35 (Tables 7.1–7.7).

Sharp force trauma is observable on the frontal bone (downward blow at an angle of 8°) and both parietals (downward blow at an angle of 25°) (Figures 7.4a and 7.4b). They were apparently inflicted by a person standing at the victim's right side, in two consecutive actions of the same hand. As both injuries could have been fatal, the beheading presumably took place shortly afterwards.

Sharp force traumas consistent with decapitation occur on the occipital bone, base of the skull and right mastoid process (Figures 7.4c–f).

Other palaeopathological changes are three antemortem blunt force injuries: of the frontal and left parietal bone (both 0.5 cm in diameter), and of the occipital bone (1 × 3 cm).

Skull No. 4 (G 23)

Skull No. 4 belonged to a male aged 20–30 (Tables 7.1–7.7).

There is sharp force trauma on the frontal bone (blow delivered from left to right by a right-handed person standing in front of the victim) and in the middle of the zygomatic arch (upward blow, leaving a nearly horizontal mark) (Figures 7.5a and 7.5b). Although severe, these injuries, inflicted in two consecutive actions of the same hand, should not have been fatal. Presumably, the decapitation was carried out shortly afterwards.

Sharp force trauma consistent with decapitation occurs on the fourth cervical vertebra (Figures 7.5c and 7.5d).

Another observable palaeopathological change is an antemortem blunt force injury (0.7 cm in diameter) of the frontal bone 2 cm from the bregma (Figure 7.5a).

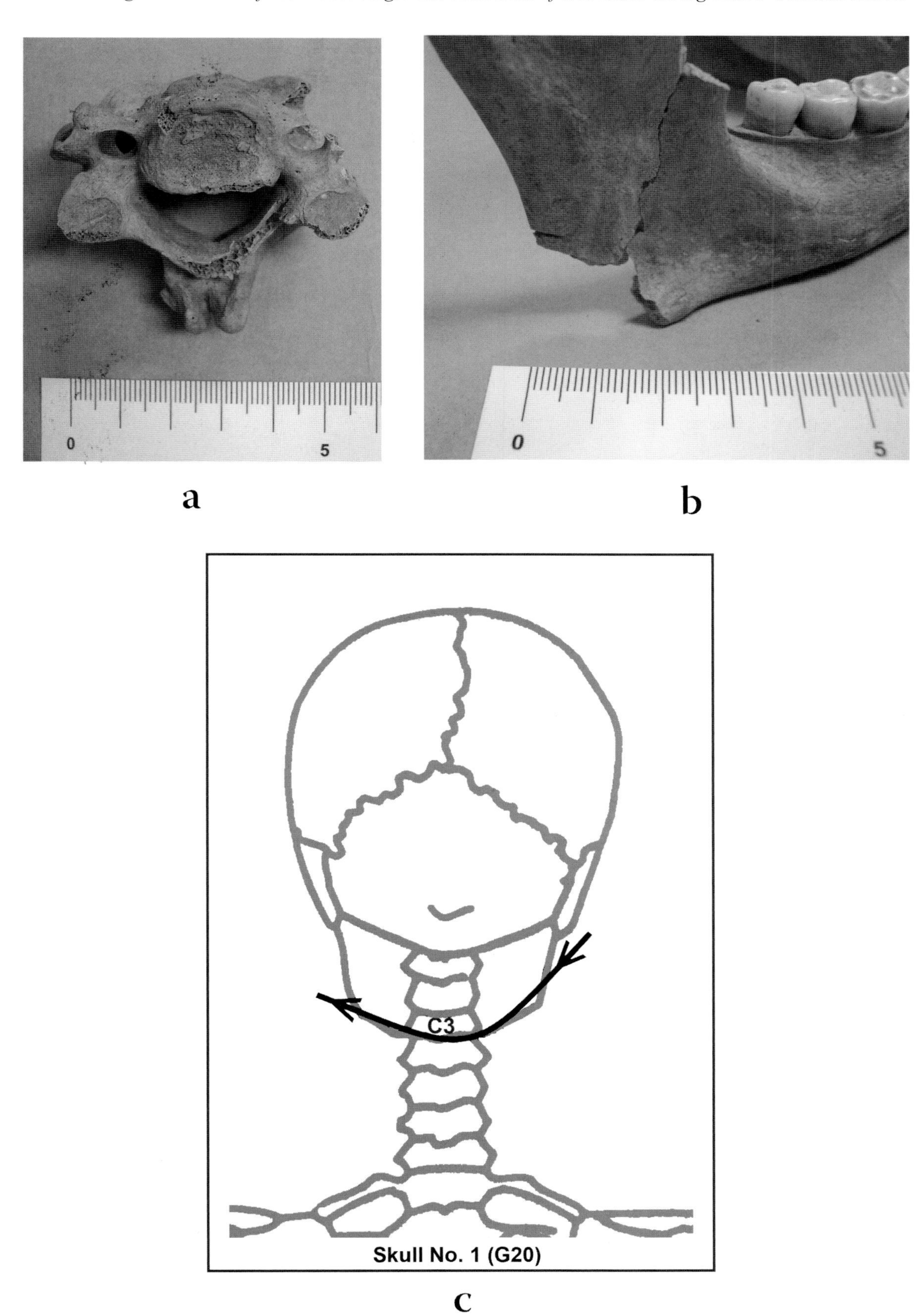

Figure 7.2. a) Sharp force trauma consistent with decapitation on third cervical vertebra. b) Sharp force trauma consistent with decapitation on right gonion. c) Reconstruction of decapitation technique.

Table 7.2 Cranial measurements and indices

CRANIAL SKELETON (cm)	G 20	G 21	G 22	G 23	G 24
Primal cranial measurements					
Maximum cranial length (g-op)	18.20	17.90	18.20	17.50	-
Maximum cranial breadth (eu-eu)	14.50	13.80	13.50	14.60	13.00
Basion/bregma height (b-ba)	-	14.80	13.40	-	-
Cranial Index	79.67	77.09	74.17	83.42	-
Cranial Length-Height Index	-	82.68	76.37	-	-
Cranial Breadth-Height Index	-	107.24	87.15	-	-
Mean Basion–Height Index	-	93.37	84.81	-	-
Cranial module	-	15.50	15.20	-	-
Porion-bregma height	11.65	11.70	12.35	11.95	9.40
Basion-porion height	-	-	-	-	-
Mean Porion-Height Index	71.25	73.81	77.91	74.45	-
Index of Flatness of the Cranial Base	-	-	-	-	-
Minimum frontal breadth (ft-ft)	9.30	9.50	-	9.20	-
Fronto-Parietal Index	64.14	68.84	-	63.01	-
Basion-prostion length	-	-	-	-	-
Basion-nasion length	-	-	-	-	-
Prognatic Index	-	-	-	-	-
Facial skeleton					
Total facial height (n-gn)	11.20	12.50	12.25	-	-
Upper facial height (n-alveolar)	6.80	7.80	7.55	-	-
Facial width or bizyg. breadth (zy-zy)	12.80	13.00	-	-	-
Total Facial Index	87.50	96.15	-	-	-
Upper Facial Index	53.12	60.00	-	-	-
Nose					
Nasal height (n-ns)	5.40	-	5.15	-	-
Nasal breadth (al-al)	2.30	2.55	2.30	2.40	2.30
Nasal Index	42.59	-	44.66	-	-
Orbits	R L	R L	R L	R L	R L
Orbital height	3.55 3.55	3.30 3.00	- 3.40	- -	- -
Orbital breadth (mf-ec)	- 3.90	- -	- 4.25	- -	- -
Orbital Index	- 91.02	- -	- 80.00	- -	- -
Maxilla					
Maxilloalveolar length (pr-alv)	-	-	-	-	-
Maxilloalveolar breadth (ecm-ecm)	-	-	-	-	-
Maxilloalveolar Index	-	-	-	-	-
Palate					
Palatal length	6.40	6.30	6.05	6.40	6.00
Palatal breadth	-	-	-	-	-
Palatal Index	-	-	-	-	-
Mandible					
Mandible lenght	10.40	11.20	11.60	10.40	10.60
Bicondylar breadth (cdl-cdl)	-	12.15	12.05	-	12.00
Bigonial breadth (go-go)	10.70	10.90	11.00	11.45	-
Height of ascending ramus	7.80	7.50	7.10	7.10	-
Minimum breadth of ascending ramus	3.20	3.25	3.05	3.20	3.15
Height of mandibular symphysis (gn-idi)	2.80	3.20	3.05	3.15	3.05
Thickness of mandibular body	0.85	1.10	1.10	1.25	0.85
Height of mandibular body	2.90	3.40	3.10	3.05	2.75
Mandibular Index	-	92.18	96.26	-	88.33
Mandibular Body Robusticity Index	29.31	32.35	35.48	40.98	30.90
Mandibular Ramus Index	41.02	43.33	42.95	45.07	-
Frontomandibular Index	86.91	87.15	-	80.35	-

Skull No. 5 (G 24)

Skull No. 5 belonged to a male aged 20–30 (Tables 7.1–7.7).

Sharp force trauma consistent with decapitation occurs on the right gonion (Figures 7.6c and 7.6d) and C2 (Figures 7.6a, 7.6b and 7.6c).

4. Discussion

Even though our analysis was limited by the fact that the postcranial skeletons were missing,[5] it was possible to verify the archaeologists' assumption about execution by beheading, and to reconstruct the individual acts of decapitation. What happened to their bodies remains

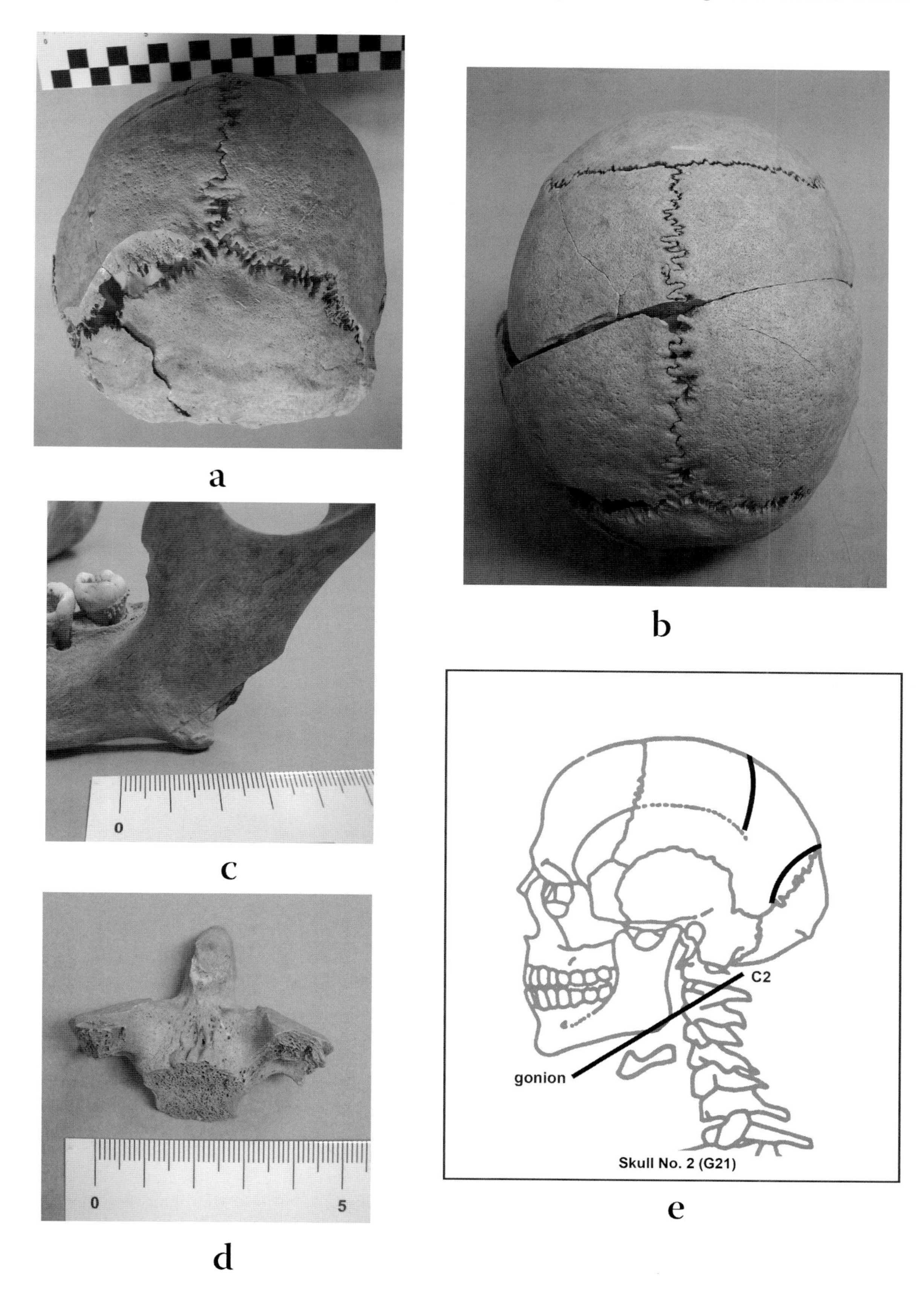

Figure 7.3. a, b) Perimortem sharp blade injuries to left parietal bone. c) Sharp force trauma consistent with decapitation on left gonion. d) Sharp force trauma consistent with decapitation on second cervical vertebra. e) Reconstruction of decapitation technique.

Table 7.3 Odontometric data

Odontometric data (diameter in cm)										
Maxilla	**G 20**		**G 21**		**G 22**		**G 23**		**G 24**	
	diameter		diameter		diameter		diameter		diameter	
Tooth	M/L	VB/L	M/L	VB/L	M/L	VB/L	M/L	VB/L	M/L	VB/L
11	postm. loss		0.90	0.80	0.80	0.75	postm. loss		0.90	0.80
12	0.60	0.60	0.70	0.70	0.65	0.60	postm. loss		postm. loss	
13	0.75	0.80	0.80	0.95	0.75	0.85	0.70	0.85	0.75	0.90
14	0.65	0.85	0.65	0.95	0.65	0.90	0.60	0.90	0.75	1.05
15	0.60	0.80	0.70	0.85	0.65	0.90	0.60	0.90	0.60	1.00
16	0.90	1.00	1.00	1.20	0.95	1.10	1.00	1.05	1.00	1.10
17	0.85	0.95	0.95	1.20	0.85	1.10	0.85	1.10	0.90	1.15
18	0.80	0.90	0.90	1.00	0.90	0.90	0.90	1.10	0.95	1.15
21	postm. loss		0.95	0.80	0.80	0.75	postm. loss		0.90	0.80
22	postm. loss		0.70	0.65	postm. loss		postm. loss		0.75	0.70
23	postm. loss		0.80	0.90	0.75	0.85	postm. loss		0.75	0.85
24	0.65	0.85	0.70	0.95	0.65	0.90	0.60	0.85	0.70	1.00
25	0.60	0.80	0.70	0.90	0.60	0.90	0.60	0.90	0.65	1.00
26	0.95	0.95	1.05	1.15	0.95	1.20	0.95	1.10	1.00	1.20
27	0.80	0.95	0.85	1.15	0.80	1.10	0.85	1.05	0.90	1.10
28	0.70	0.95	0.80	1.10	0.85	1.05	postm. loss		0.90	1.15
Odontometric data (diameter in cm)										
Mandibula	**G 20**		**G 21**		**G 22**		**G 23**		**G 24**	
	diameter		diameter		diameter		diameter		diameter	
Tooth	M/L	VB/L	M/L	VB/L	M/L	VB/L	M/L	VB/L	M/L	VB/L
31	0.50	0.60	0.55	0.65	0.50	0.55	0.50	0.60	0.55	0.70
32	0.55	0.70	0.60	0.75	0.55	0.65	0.55	0.65	0.65	0.60
33	0.60	0.80	0.65	0.75	0.65	0.80	0.60	0.80	0.70	0.80
34	0.65	0.65	0.70	0.75	postm. loss		0.65	0.70	0.70	0.80
35	0.65	0.75	0.75	0.85	0.70	0.80	0.65	0.70	0.70	0.85
36	0.95	0.90	1.10	1.05	1.00	1.00	1.05	1.00	1.05	1.05
37	0.95	0.95	1.10	caries	0.95	0.95	1.05	1.00	0.95	1.00
38	0.95	0.90	1.10	1.10	0.95	0.90	1.00	1.00	1.00	1.05
41	root		0.60	0.70	0.50	0.55	0.50	0.55	0.55	0.65
42	0.55	0.65	0.60	0.75	0.55	0.65	0.60	0.65	0.60	0.65
43	0.60	0.75	0.75	0.85	0.65	0.80	0.65	0.75	0.70	0.80
44	0.60	0.65	0.65	0.75	0.65	0.75	0.70	0.70	0.70	0.80
45	0.60	0.75	0.75	0.90	0.65	0.80	0.65	0.75	0.65	0.90
46	0.90	0.90	caries		1.00	1.00	root		1.00	1.05
47	0.85	0.90	antem. loss		0.95	0.95	1.00	1.00	1.05	1.00
48	0.90	0.85	caries		1.00	0.95	1.00	0.95	1.00	1.00

unknown; they were probably buried or disposed of elsewhere. Some other important questions also remain unanswered: were the victims restrained, and were the fatal injuries of victims G 21 and G 22 inflicted only to the cranial skeleton, in what order were they executed, etc.?

The sharp force injuries to the skulls are very similar to cut marks left by swords (Lewis, 2008). In our view, however, such entrance and exit angles can only be produced by a bladed weapon similar to the sword but slightly curved and thinner and therefore more manoeuvrable (Figures 7.2c, 7.3e, 7.4f, 7.5d and 7.6d); in other words, the sabre. Incidentally, the sabre was a widely used weapon in the Balkans from the 16th through the 20th century (Šercer, 1979; Milosavljević, 1993).

Victim G 20 was probably kneeling, his back to the executioner and facing the 'public'. If so, the executioner must have been right-handed (Figure 7.2c). The sabre went downward, nicked the right gonion at an angle of 55° and in its upward arc sliced the lower left part of C3 at an angle of 30°.

Victim G 21 had suffered two lethal injuries to the left parietal bone from a sharp-edged weapon, probably a sabre, by a person standing above him or at his left side. The victim was either killed or knocked unconscious, and thus his head had to be laid down on a block, on its right side, and was severed with a single blow (Figure 7.3e).

Victim G 22 had also sustained two fatal injuries, one to the frontal, the other to both parietal bones, and was either killed instantly or lost consciousness. Hence, as in the previous case, his head was propped up on a support, sideways on the left cheek, and severed probably using the usual technique. What is intriguing in this particular case are perimortem sharp blade injuries on the occipital bone inflicted immediately after the beheading (first blow at an angle of 45°, second at 10°, and third, affecting almost the entire base of the skull, at an angle less than 10°). The

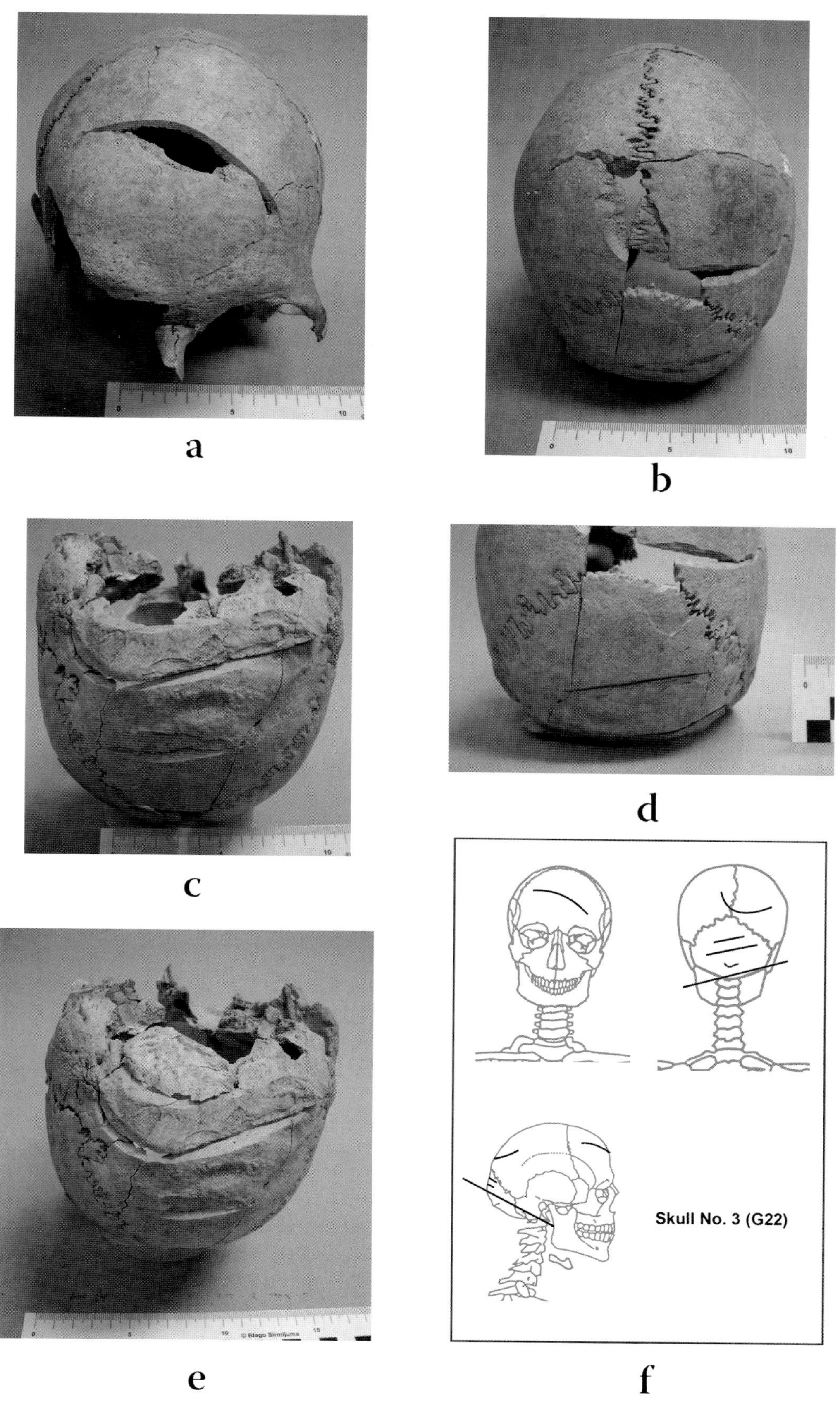

Figure 7.4. a) Perimortem sharp blade injury to frontal bone. b) Perimortem sharp blade injury to both parietals. c, d, e) Few perimortem sharp blade injuries consistent with decapitation to occipital bone. f) Reconstruction of decapitation technique.

Table 7.4 Caries

Maxilla Tooth	Caries (diameter in cm) G 20	G 21	G 22	G 23	G 24
11	-	-	-	-	-
12	-	-	-	-	-
13	-	-	-	-	-
14	-	-	-	-	-
15	-	-	-	-	-
16	-	spot (L)	-	-	-
17	-	-	-	-	-
18	-	3 spots (O)	-	-	-
21	-	-	-	-	-
22	-	-	-	-	-
23	-	-	-	-	-
24	-	-	-	-	-
25	-	spot (O)	-	-	-
26	-	spot (L)	-	-	-
27	0.25 (D)[1]	-	-	-	-
28	5 spots (O) and 0.30 (M)	spot (O)	spot (B)	-	-
Mandibula Tooth	**Caries (diameter in cm) G 20**	**G 21**	**G 22**	**G 23**	**G 24**
31	-	-	-	-	-
32	-	-	-	-	-
33	-	-	-	-	-
34	-	-	-	-	-
35	-	-	-	-	-
36	-	2 spots (O)	-	-	-
37	-	0.80 (gross; O/B)	-	-	-
38	-	0.30 (O)	-	-	-
41	-	-	-	-	-
42	-	-	-	-	-
43	-	-	-	-	-
44	-	-	-	-	-
45	-	-	-	-	-
46	0.30 (D)	gross gross	-	-	-
47	0.30 (M)	-	-	-	-
48	-	gross-gross	-	-	-

[1] All caries lesions are on the crowns (O – occipital; M – mesial; D – distal; L – lingual; B – buccal), except, of course, gross caries (O/B – occluso-buccal) and gross-gross caries. 'Gross caries' is the term used to describe a lesion that has grown to the point that it includes several possible sites of initiation, and, therefore, its original site cannot be determined. 'Gross gross' carious cavity, involving the loss of so much of the tooth that it is not possible to determine whether the lesion was initiated in the crown or root, and there is a clear opening into an exposed pulp chamber or root canal (Hillson 2001).

Table 7.5 Dental diseases

Dental diseases present	G 20	G 21	G 22	G 23	G 24
Maxilla					
Hypoplasia[1]	medium	considerable	medium	slight→medium	medium
Periodont. disease	medium	medium→consider.	medium→consider.	medium→consider.	slight
Calculus	slight	medium→consider.	medium→consider.	slight→medium	-
Periapical abscesses	-	-	-	-	-
Mandibula					
Hypoplasia	medium	considerable	medium	slight→medium	medium
Periodont. disease	medium	medium→consider.	medium→consider.	medium→consider.	slight
Calculus	medium	medium→consider.	slight	slight→medium	medium
Chronical periapical abscesses	-	37 (B: 0.4 cm) and 46 (B: 0.6 cm)	-	46 (B: 0.5 cm)	-

[1] Scoring for hypoplasia, periodontal disease and calculus is taken from Brothwell (1981: 155 and 156).

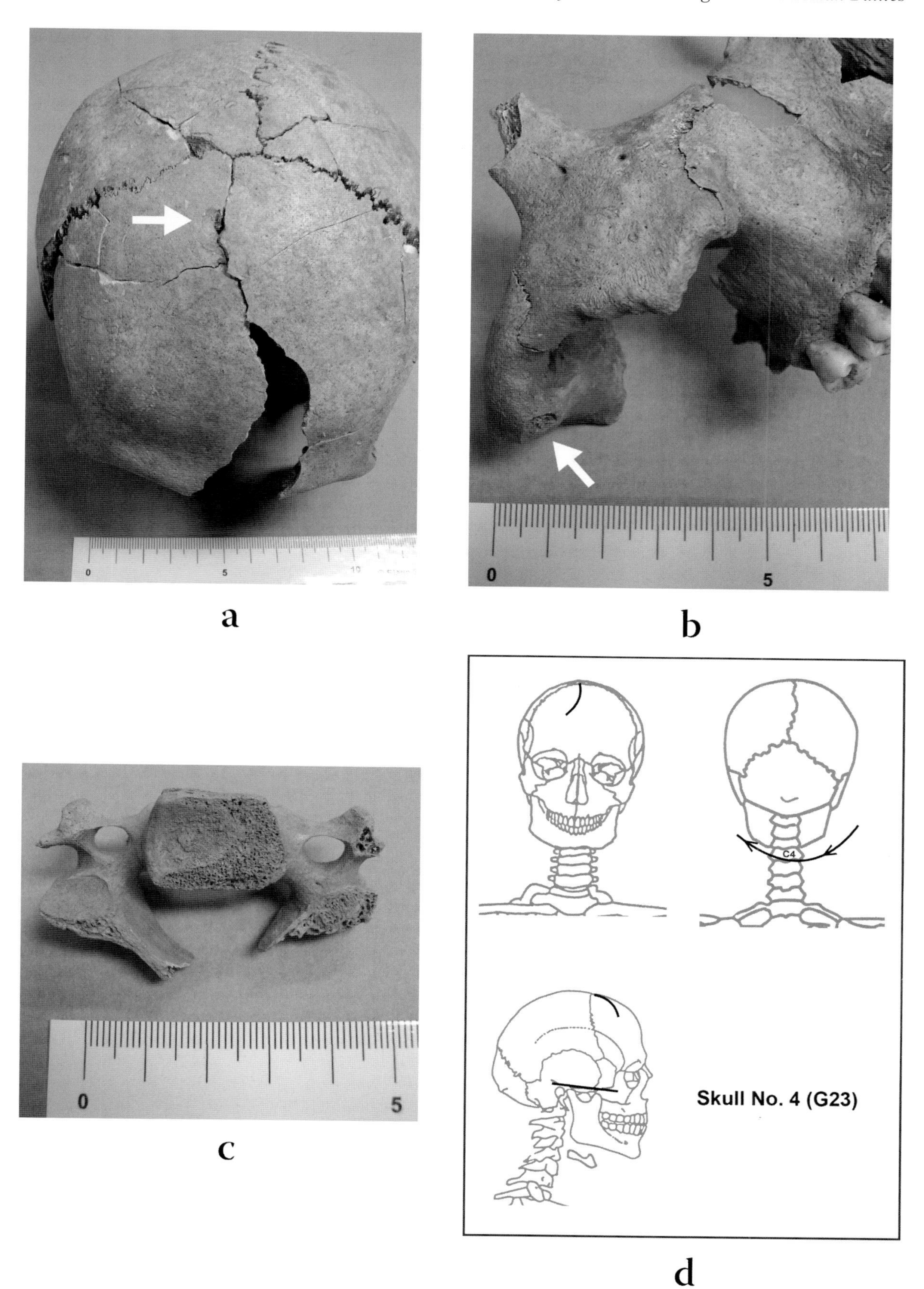

Figure 7.5. a) Blunt force injury to frontal bone. b) Perimortem sharp blade injury to zygomatic arch. c) Sharp force trauma consistent with decapitation on fourth cervical vertebra. d) Reconstruction of decapitation technique.

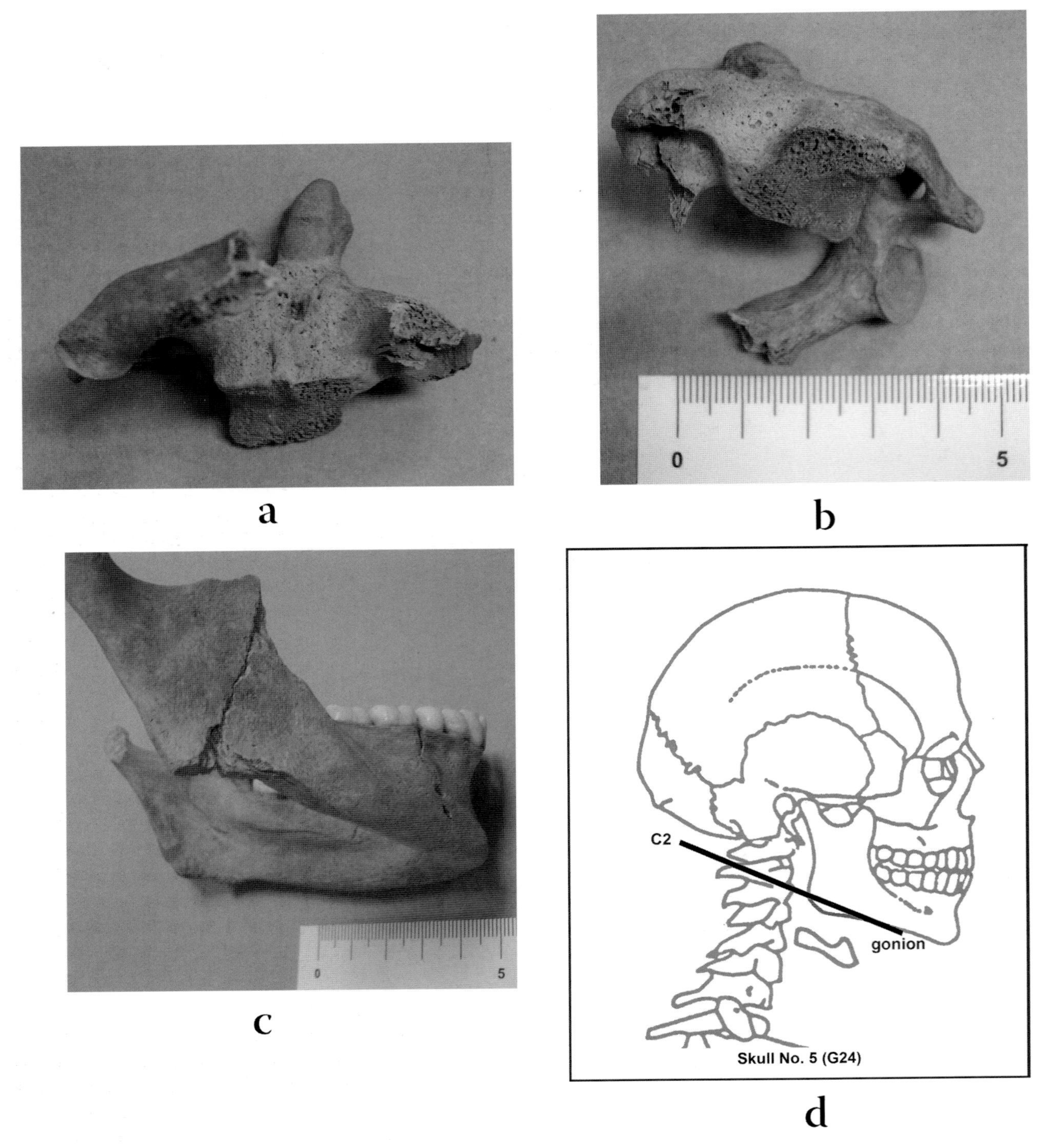

Figure 7.6. a, b) Sharp force trauma consistent with decapitation on second cervical vertebra. c) Sharp force trauma consistent with decapitation on right gonion. d) Reconstruction of decapitation technique.

Table 7.6 Anomalies of teeth and occlusion

Anomalies of teeth and occlusion present	**G 20**	**G 21**	**G 22**	**G 23**	**G 24**
Anomalies of teeth					
Maxilla					
Rotation of teeth	14, 15, 24 and 26	-	-	-	-
Mandible					
Rotation of teeth	-	-	41	-	-
Occlusion	edge-to-edge	protrusion	edge-to-edge	edge-to-edge	protrusion

Table 7.7 Non-metric variations

Non-metric variations[1] present	**G 20**	**G 21**	**G 22**	**G 23**	**G 24**
Sulci frontales	one on the right side	one on the right side	-	-	-
Foramen infraorbitale accesorium	-	three on the left side	two on the left side	-	-
Ossa suturae lambdoideae	-	-	one on the right and two on the left side	two on each side (d= 0.5 × 0.8 cm)	-
Linea nuchae suprema	prominent	-	-	prominent	-
Foramen zygomaticofaciale	two on each side	-	two on the left side	two on the right side	two on the left side
Trace expression of the squamo-mastoid suture	on both sides	-	-	-	-

[1] Observation of non-metric variations is according to Hauser and De Stefano (1989).

purpose of these blows might have been to make it 'easier' to set the head upon a stake for public display. This is the only skull that shows evidence of further violent penetration of the base of the skull when impaled (Figures 7.4c and 7.4e). The head may not have been displayed for very long, considering that the mandible and the fragment of the atlas (probably somehow left stuck in the base of the skull) remained attached to the head with the surrounding soft tissues.

Victim G 23 was beheaded in the 'regular' way. The injuries inflicted with a sharp blade prior to beheading were not fatal. The person was probably on his knees, with the executioner standing behind him. The sabre, swung in a semi-arc from right to left, affected only the lower left side of the body (at an angle of 10°) and inferior articular process of the fourth cervical vertebra. The force of the blow sliced off the *arcus vertebrae* and *processus spinalis* (Figure 7.5d).

Victim G 24 was probably kneeling, facing the 'public', while the executioner, undoubtedly right-handed, was standing behind him (Figure 7.6d). The sabre was swung downward, first cutting the right gonion at an angle of 10°, and then the lower right part of C2 at an angle of 20°. Considering the markedly high decapitation location, the head may have also been propped up sideways on its left cheek.

5. Conclusion

The nature of the find, above all the severed heads, and the find-spot lead us to think of a punishment imposed for wartime crimes, such as mutiny, espionage or treason. Such an assumption is suggested by the numerous perimortem injuries probably sustained in a head-to-head battle that had preceded the beheading, and by the public display of one of the heads. The large number and severity of injuries in one case (No. 3) suggests that this individual may have been the prime culprit, possibly the leader of a mutiny, as additionally corroborated by the fact that this is the only head that was publicly displayed. It seems reasonable to assume that the execution of the five persons took place during the Austro-Ottoman wars of the late 17th century, most likely during the Ottoman recapture of Belgrade in 1690.

The heads were probably disposed of shortly after the execution, as evidenced by the presence of mandibles and anatomic connection of cervical vertebrae as yet unaffected by postmortem decomposition. Moreover, they were neatly arranged, which suggests a measure of respect for the executed. The form of punishment suggests distinguished civilians or high-ranking militaries. That they indeed were notable and well off is indicated by their healthy and well-kept teeth (no lifetime tooth loss). Dental analyses have shown very low levels of tooth wear (even considering their relatively young adult age) and caries. The only exception is skull No. 2 with a few carious teeth and one tooth lost antemortem (probably also due to caries). All this indicates well-fed individuals and a diet rich in proteins of animal origin (Larsen, 1999: 76–77).

The skeletons discovered in the immediate vicinity, however, call for a different interpretation. The absence of burial pits, and the orientation and partial intertwinement of the bodies suggest that they were simply dumped in the ground, possibly out of a cart; which was not an uncommon treatment of the urban poor, as evidenced by recent archaeological excavations in Sremska Mitrovica (Miladinović-Radmilović, 2011a: 138–177, 2011b). Hence, the 'burial' of noblemen, or rather the disposal of their remains (bodiless heads), with the urban poor would not have merely been an act of denigration commensurate with the crime they had been punished for, but the effacement of the fact that they had ever existed; a sad wartime occurrence in all times and places in human history.

Acknowledgements

This paper is a result of the project *Urbanization processes and development of medieval society* (No. 177021) funded by the Ministry of Education, Science and Technological

Development of the Republic of Serbia. We express our gratitude to Mr Miro Radmilović for the reconstruction drawings of the beheadings made according to our observations, to our colleague Neda Dimovski (Municipal Museum, Subotica) for her expert assistance, to the student Dragica Bizjak for assistance in preparing the osteological material for analysis and useful suggestions, and to Marina Adamović Kulenović for translating the text.

Notes

1. Information contained in the excavation records of the Institute of Archaeology, Belgrade/Belgrade Fortress Research Project.
2. We were not in a position to carry out AMS dating because these skulls are unique archaeological samples and therefore the preparation of bone sections was not allowed.
3. It is known that this method is not population specific and this must be taken into consideration during the age estimation of different populations particularly in archaeological investigations. Since the preparation of bone sections was not allowed, we could not use the method of root dentine translucency for age estimation. On the other hand, pulp/tooth area ratio as an indicator of age is quite a new method for age estimation, which requires a radiographic image of an examined tooth, computer-aided drafting software and trained personnel. Unfortunately, this method was too expensive and we were not able to perform such age assessments.
4. Regrettably, as can be seen in Figure 7.1b, only the bones of a single skull (No. 1) are not deformed by soil pressure.
5. Apart from a few cervical vertebrae.

References

Anderson, T. 2001. Two decapitations from Roman Towcester. *International Journal of Osteoarchaeology* 11, 400–405.

Bass, W.M. 1995. *Human Osteology: A Laboratory and Field Manual*. Missouri Archaeological Society, Columbia, MO.

Brothwell, D.R. 1981. *Digging Up Bones: The Excavation, Treatment and Study of Human Skeletal Remains,* 3rd edition. British Museum (Natural History), London and Oxford University Press, Oxford.

Buikstra, J.E., Ubelaker, D.H. 1994. *Standards for Data Collection from Human Skeletal Remains*. Arkansas Archeological Survey Research Series No. 44, Fayetteville.

Daniell, Ch. 1997. *Death and Burial in Medieval England 1066–1550*. Routledge, London and New York.

Ferembach, D., Schwidetzky, I., Stloukal, M. 1980. Recommendations for age and sex diagnosis of skeletons. *Journal of Human Evolution* 7, 517–549.

Harman, M., Molleson, T.I., Price, J.L. 1981. Burials, bodies and beheading in Romano-British and Anglo-Saxon cemeteries. *Bulletin of the British Museum of Natural History – Geology* 35(3), 145–188.

Hauser, G., De Stefano, G.F. 1989. *Epigenetic Variants of the Human Skull*. E. Schweizerbart'sche Verlagsbuchhandlung, Stuttgart.

Hillson, S. 1990. *Teeth*. Cambridge University Press, Cambridge.

Hillson, S. 1996. *Dental Anthropology*. Cambridge University Press, Cambridge.

Hillson, S. 2001. Recording dental caries in archaeological human remains. *International Journal of Osteoarchaeology* 11, 249–289.

Larsen, C.S. 1999. *Bioarchaeology: Interpreting Behavior from the Human Skeleton*. Cambridge University Press, Cambridge.

Leben des Herren Baron Doxat von Morez 1757 – Leben des Kaiserlichen Generals und berühmten Ingenieurs Herren Baron Doxat von Morez, Welcher Anno 1738. den 20sten Martii zu Belgrad enthauptet worden, nebst besondern Umständen des damaligen Türken-Krieges, 1757/ Život carskog generala i slavnog inženjera gospodina barona Doksata de Moreza pogubljenog 20. marta 1739. u Beogradu, uz opis pojedinih dešavanja u tadašnjem ratu protiv Turaka, 1757. Biblioteka grada Beograda, Belgrade 2006.

Levental, Z. 1989. *Britanski putnici u našim krajevima od sredine XV do početka XIX veka*. Dečje novine, Gornji Milanovac.

Lewis, J.E. 2008. Identifying sword marks on bone: criteria for distinguishing between cut marks made by different classes of bladed weapons. *Journal of Archaeological Science* 35, 2001–2008.

Lovejoy, C.O. 1985. Dental wear in the Libben Population: Its functional patterns and role in the determination of adult skeletal age at death. *American Journal of Physical Anthropology* 68, 47–56.

Miladinović-Radmilović, N. 2011a. *Sirmium – Necropolis*. Arheološki institute, Belgrade and Blago Sirmijuma, Sremska Mitrovica.

Miladinović-Radmilović, N. 2011b. Exostoses of the external auditory canal. *Starinar* LX/2010, 137–146.

Milosavljević, B. 1993. *Mačevi u zbirkama Vojnog muzeja XIV–XX veka*. Vojni muzej, Beograd.

Popović, M. 2006. *Beogradska tvrđava*, 2nd revised edition. Javno preduzeće Beogradska tvrđava, Belgrade.

Šercer, M. 1979. *Sablje*. Povijesni muzej Hrvatske, Zagreb.

Škalamera, Ž. 1973a. Planovi Beograda iz 1688. godine. *Urbanizam Beograda* 20, 15–18.

Škalamera, Ž. 1973b. Planovi barokne rekonstrukcije Beograda iz 1717–1740. godine. *Urbanizam Beograda* 23, 13–18.

Škalamera, Ž. 1973c. Planovi Beograda iz 1789. godine. *Urbanizam Beograda* 24, 19–24.

Škalamera, Ž. 1975a. Područje Knez Mihailove ulice u osmanlijskom šeheru Beogradu od 1521. do 1688. godine, in: *Knez Mihailova ulica, zaštita nasleđa, uređenje prostora*. Belgrade, Galerija Srpske akademije nauka i umetnosti 26, pp. 18–21.

Škalamera, Ž. 1975b. Područje Knez Mihailove ulice od 1688. do 1717. godine, in: *Knez Mihailova ulica, zaštita nasleđa, uređenje prostora*. Galerija Srpske akademije nauka i umetnosti 26, Belgrade, p. 22.

Škalamera, Ž. 1975c. Područje Knez Mihailove ulice u doba barokne rekonstrukcije Beograda 1718–1740. godine, in: *Knez Mihailova ulica, zaštita nasleđa, uređenje prostora*. Galerija Srpske akademije nauka i umetnosti 26, Belgrade, pp. 23–27.

Škalamera, Ž. 1975d. Područje Knez Mihailove ulice u drugoj polovini XVIII veka i početkom XIX veka, in: *Knez Mihailova ulica, zaštita nasleđa, uređenje prostora*. Galerija Srpske akademije nauka i umetnosti 26, Belgrade, pp. 28–31.

Škalamera, Ž. 1975e. Područje Knez Mihailove ulice za vreme Prvog srpskog ustanka, in: *Knez Mihailova ulica, zaštita nasleđa, uređenje prostora*. Galerija Srpske akademije nauka i umetnosti 26, Belgrade, p. 32.

Škalamera, Ž. 1975f. Područje Knez Mihailove ulice u prvoj fazi razvoja prestonice Kneževine Srbije od 1815. do 1867.

godine, in: *Knez Mihailova ulica, zaštita nasleđa, uređenje prostora*. Galerija Srpske akademije nauka i umetnosti 26, Beograd, pp. 33–39.

Walker, P. L. 2001. A bioarchaeological perspective on the history of violence. *Annual Review of Anthropology* 30, 573–596.

Wells, C. 1982. The human remains, in: McWhirr, A., Viner, I., Wells, C. *Romano-British Cemeteries at Cirencester*. Cirencester Excavation Committee, Cirencester, pp. 135–202.

Wiltschke-Schrotta, K. and Stadler, P. 2005. Beheading in Avar times (630–800 A.D.). *Acta Medica Lituanica* 12(1), 58–64.

8. The Remains of a Humanitarian Legacy: Bioarchaeological Reflections of the Anatomized Human Skeletal Assemblage from the Worcester Royal Infirmary

A. Gaynor Western

In 2010, archaeological excavations carried out in the grounds of the Worcester Royal Infirmary, Worcestershire, England, led to the discovery of disarticulated human remains, some of which displayed clear evidence of anatomization. The Infirmary served as a voluntary hospital during the late Georgian and Victorian period to provide access to health care for the poor under the direction of Sir Charles Hastings, founder of the British Medical Association. Such hospitals also formed centres for medical education and research, not only for observing diseases and developing treatments but also facilitating access to bodies for post-mortem examinations. The skeletal remains examined from the site exhibited saw and cut marks that represented bisection, amputation, trepanation, craniotomy and dissection. This paper seeks to explore the circumstances in which modified human remains came to be discarded in non-normative deposits and how this reflects contemporary attitudes to the body and body parts in the Industrial period. The findings illustrate the importance of placing osteological observations within their archaeological context when interpreting such hospital assemblages. This combination of bioarchaeological evidence provides a unique insight into medical practice during this period.

Keywords: Worcester; Charles Hastings; Amputation; Dissection; Hospital; Post-mortem

> 'Some people die at 25 and aren't buried until 75'.
> Benjamin Franklin

1. Introduction

In 2010, archaeological excavations carried out in the grounds of the Worcester Royal Infirmary led to the discovery of four discrete deposits of disarticulated human remains. In total, 1828 fragments were recovered, 236 of which displayed clear evidence of modification, having been bisected by saw or defleshed by knife. Modified elements included the mandible, sphenoid, cranium, vertebrae, clavicle, pelvis, long bones, foot and a small number of ribs. The human remains consisted of adult males and females as well as sub-adults. The find was unexpected. Preliminary documentary research had indicated that no burial ground was associated with the Infirmary. The somewhat startling discovery prompted questions as to the circumstances in which the human bones had been interred at the Infirmary and who these remains represented.

2. Voluntary Hospitals, Dissection and the Humanitarian Movement

Worcester Infirmary was established on the open grounds of the Castle Street site in 1771 and represented a new flagship in the philanthropic endeavours of the city of Worcester. Architecturally, the plan of the building was almost identical to that of a typical metropolitan hospital, its design aimed at providing light and airy wards in which the needy and ailing poor could be treated with the latest medical care (McMenemey, 1947). The late Georgian and early Victorian periods saw exponential changes in society due to scientific and technological advances but these were coupled with population growth, increasing overcrowding and disease rife amongst the poorer, working classes who had no independent means to support themselves should they become ill and unable to work. The voluntary hospitals (also known as infirmaries) allowed access for the first time to professional health care for the poor. The concept of health at this time was undergoing a period of redefinition and the age saw major ontological shifts in beliefs about the body. An increasingly somatic view of health became prevalent and disease was now understood within the medical discipline as a corruption of an *a priori* state of health that could be avoided or treated, rather than a divine manifestation of spiritual ills (Binski 1996; Risse, 2010).

The new recognition of human ability to intercede in cases of suffering gave rise to the humanitarian movement during the Enlightenment period, with an unprecedented number of charitable institutions being established to provide social welfare and education to those who could not afford it. Given the concomitant naturalistic shift in ontology, the body and its corruption became the locus of the new humanitarian rhetoric (Laqueur, 1989) grounded in secularism. Dirt and disease came to represent to many a state of ignorance and the social ills of the time. For example, it was during the rise of the voluntary hospitals that 'occupational disease' was first recognised and medical practitioners observed that unsanitary working conditions could directly contribute to illness (Bartrip, 2003). With mechanisation in its infancy, the physical functioning of the working class population was imperative to the success of industry. The philanthropic intervention of doctors in administering aid to the sick poor via the voluntary hospital institution was often justified by aiming to render an individual fit for the labour force as part of social progressivism (see Liston, 1842; Brown 2011). Disease aetiology, however, was in its earliest stages of discovery and treatment at the Infirmary during this period was crude and experimental. Generally, it consisted of the administration of medicines such as spirit of nitrous aether, camphor mixture, saline antimonial mixture, port wine, carbonate of soda, mercury with chalk, compound chalk powder, calomel and opium (Hastings, 1841). In addition, blood-letting was also practised; heads were shaved and cupped, and leeches were used (Hastings, 1841). Surgical interventions were severely restricted at this time and the only major operations carried out that would leave traces on the skeleton were trepanation and amputation.

Whilst the public face of the hospitals was to provide facilities for curing disease, the infirmaries also functioned to provide experience and data for the medical practitioners, with hospital wards serving as platforms for bodily observation. Observations of patients included regular inspection of the mouth (tongue), chest, abdomen, bowels, skin and pulse (Hastings, 1841). Biological reductionism dominated medical epistemology. In order to understand symptoms, the new anatomico-pathological medical model (Foucault, 2003), espoused by pioneers such as Laennec and Bichat in France (Brown, 2011), demanded that these external signs of illness were correlated with observations of internal lesions and organs. Given the lack of anaesthesia until the 1840s, this could only be achieved after death via the post-mortem examination. An increasingly utilitarian Victorian cultural attitude permeated the administration of medical charity and it became a non-negotiable contract between the eligible individual and the institution. Very often body parts were deemed the property of the hospital as part of the agreement of admitting patients (Edwards and Edwards, 1959; Fowler and Powers, 2012); a repugnant but inescapable consequence for the poor accepting institutional medical aid. The act of post-mortem examination and dissection was a major transgression of norms of the period. Even during life, a physical examination of a living patient, especially of females, broke conventional social proprieties (Reiser, 1978). After death, Christian doctrine dictated that the body was to be interred complete for the Resurrection (Richardson, 1988; Strange 2005; Cherryson *et al.*, 2011). Breaking open the body for observation of internal structures and lesions crossed both a physical and social boundary that for many was unacceptable on moral and religious grounds but nonetheless, during this period it was a requisite process for understanding disease aetiology. To this end, Worcester Infirmary was provided with a dissection room at its inception (McMenemey, 1947).

Post-mortem examinations were regularly undertaken in the era preceding the Anatomy Act 1832. In its very early history, the autopsy tended to be carried out where the cause of death was required to be established for legal reasons (Lane, 2001), though post-mortem examinations were also undertaken at the request of next-of-kin (Crossland, 2009) or at the persuasion of the attending physician (Manuel, 1996). The body and any evidence pertaining to it were required to be openly viewed by both the Coroner and jury (Burney, 1994). As the medical expert became more frequently called upon, the post-mortem became a cornerstone in forensic medicine, culminating in the Medical Witnessing Act 1836 (Hurren, 2009). As with therapeutic treatment, biological reductionism was a principal element in ensuring the privilege of and authorisation for medical experts to intervene after death and to modify corpses (see Burney, 1994). In private medical practice, post-mortem examinations were undertaken on individuals from all social classes (Chaplin, 2012). Indeed, bioarchaeological evidence of post-mortem examination in Worcester has been recorded in one individual from St. Andrew's (Western, 2006), one of the poorest parishes, as well as in one middle class individual interred in a burial vault at Tallow Hill cemetery (Ogden, 2003). At the Infirmary, post-mortems would have been carried out on deceased in-patients, patients admitted as emergencies as well as on those who had died in unknown, sudden or accidental circumstances, as requested by the Coroner extramurally. Since permission had to be obtained from the relatives of the deceased, early post-mortems could be restricted to specified anatomical areas (see, for example, Peacock, 1860). There was no standardised procedure by which to carry out a post-mortem examination at this time and adversarial legal process often exposed medical observations to be weak, empirical and inconsistent (Hamlin, 1986).

Medical practice was dominated by practitioners working on an individual basis, even within establishments such as voluntary hospitals (McMeneney, 1947), and this is reflected in post-mortem procedures and reports that very often emphasise the area of interest or expertise of the practitioner. This is particularly evident in cases involving the death of in-patients, where dissection was undertaken in particular to correlate morbid lesions with symptoms observed on the ward (Hastings, 1827, 1855). Morbid dissection became a particularly prominent practice in the 19th century when the study of pathology, in part based upon specimen collections, became an essential component of medical education. In essence, the post-mortem was an opportunity not only to establish the cause of death but also to study and collect pathological specimens (see Liston 1842; Alberti 2011). Prior to the Anatomy Act 1832 and the Medical Witnessing Act 1836, 'opening the body' (Chaplin, 2012) to establish the cause of death was also known as 'dissection' (Hastings, 1827; Lane, 2001) and there was no strict distinction between the practice of morbid anatomy and autopsy. Even after the lawful Acts when the term 'post-mortem' was introduced, the methodology for dissection for the purposes of establishing the cause of death followed the general approach used for anatomical dissection, until Virchow's standardised procedures for autopsy were published in English in 1880.

In addition to post-mortem examinations for the purposes of morbid anatomy and autopsy, anatomical dissection was also practiced at the Infirmary on executed murderers and after the Anatomy Act 1832, on unclaimed cadavers, usually those of in-patients or those from the workhouse or prison (Richardson, 1988). The allowance by the Act of the dissection of unclaimed cadavers without obtaining the consent of the deceased was deemed to be necessary on humanitarian grounds for the greater good of society and also served to circumvent the atrocity of illegal grave-robbing. The methodological approach taken towards anatomical dissection was reasonably standardised, following dissection manuals such as Holden's (1851). It was generally more destructive than morbid anatomy, consisting of a systematic stripping of the soft tissues from all parts of the body (Cherryson, 2010) as well as the severing of body parts for sharing between students (Parsons, 1831) or even institutions (Hurren, 2004). Cadavers were also used for the practise of surgical procedures, such as amputation and trepanation (Carden, 1864).

The Anatomy Act 1832 was designed to facilitate access of medical practitioners to cadavers for anatomical dissection whilst granting family or friends the rights to intervene on behalf of the deceased to prevent profligate treatment of the corpse. A condition of the Act stated that dissected unclaimed bodies were to receive a decent burial according to Christian custom, paid for by the institution. However, body parts were not specifically mentioned in the Act and their treatment was a grey area in law (Richardson, 1988). In practice, body parts could be removed and retained with no legal repercussions, whether for anatomical or pathological study. Only in 1961 with the supplementation of the Human Tissue Act was consent of the individual to be obtained prior to dissection of their body. Legislation relating specifically to the public use of human body parts less than 100 years old for research, educational or display purposes was later drawn up in the Human Tissue Act 2004.

Many hospitals established cemeteries within their own grounds for the burial of in-patients and dissected bodies, such as the Newcastle Royal Infirmary (Chamberlain, 2012) and the Royal London Hospital (Fowler and Powers, 2012), although some purchased burial plots in nearby municipal cemeteries, i.e. Edinburgh (Henderson *et al.*, 1996) and Cambridge Infirmaries (Hurren, 2004). No burial ground existed at Worcester Infirmary; documentary evidence shows that coffins were purchased and it is likely that bodies were transported from the Infirmary to the local city cemetery at Tallow Hill or returned to the parish of origin for burial. Bioarchaeological analysis of the human remains unexpectedly discovered in the grounds of the Worcester Royal Infirmary aimed to elucidate the motives behind the non-normative deposits and to help identify the origins of the remains. While the historical background of the Infirmary and its practitioners have been discussed in detail previously (see McMenemey 1947; Western 2012), the observations made of the human skeletal assemblage presented here provide the only tangible evidence of medical practice at the Infirmary since no admissions registers or case records for the hospital survive.

3. Bioarchaeological Evidence

Excavations within the grounds of the Worcester Infirmary revealed four deposits of disarticulated human remains, as illustrated in Figure 8.1 (see Western and Kausmally, 2011). The precise date of the deposits is unknown but the associated pottery and glass assemblage suggests a broad date from the late 18th to 19th century, with a number of finds suggesting a likely date of the earlier part of the 19th century (Evans, in prep.). Stratigraphically, the deposits pre-date 1930 and respect the first phase of the hospital building. They are, therefore, considered to be 'post-medieval'. Three contexts were contained within discrete pits. Two of the deposits [5003] and [10003] were contained in shallow pits [5004] and [10004] respectively. The majority of the remains were excavated from pit [5004], which although shallow was sizeable and measured 2m × 2.25m. In comparison, pit [10004] was smaller, measuring 1.15m in width. The third context [5008] consisted of one fill within a large, deep pit [5013]. This large pit and the smaller pit containing context [5003] were located adjacent to the walls of the first phase of the Infirmary building. Context [10003], contained in the second small

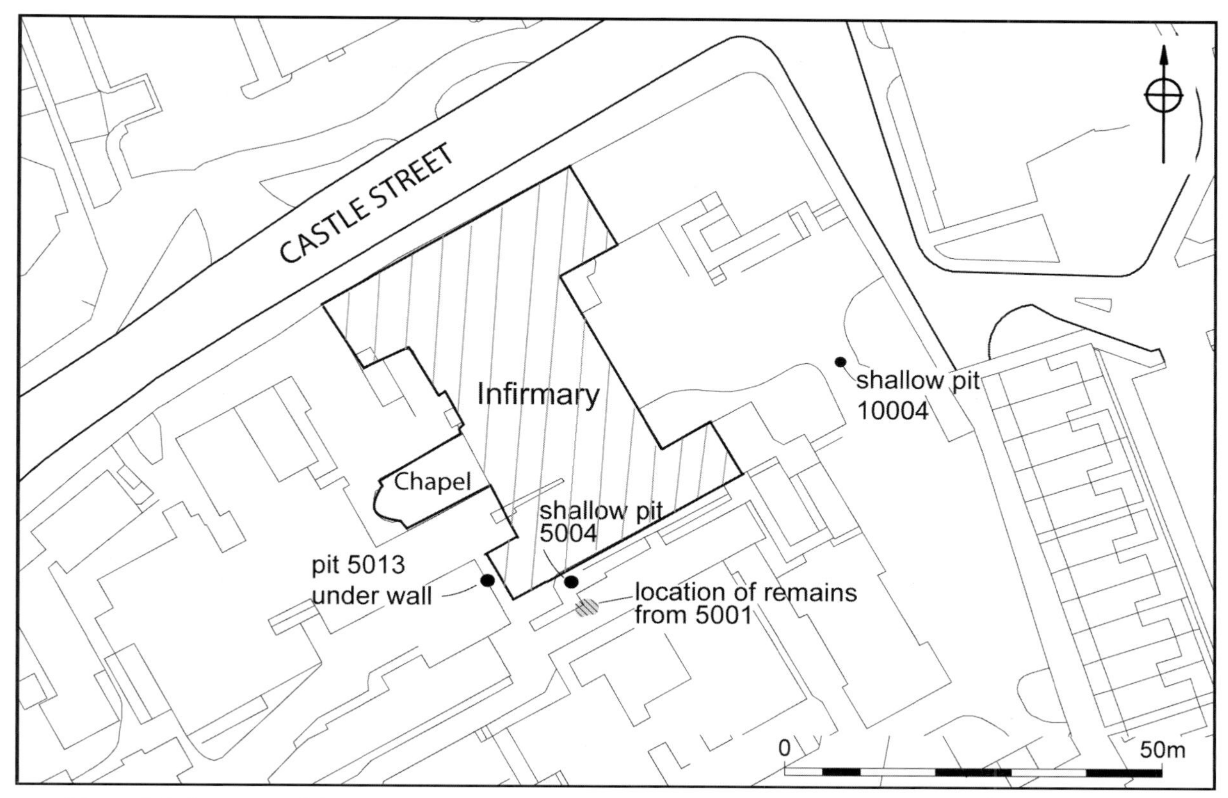

Figure 8.1. Location of the pits containing human remains in the grounds of the Worcester Royal Infirmary.

pit was situated in the front garden, approximately 20m distant to the hospital. One additional small deposit [5001] was heavily disturbed and its original context could not be ascertained. Only limited areas of the grounds were excavated and it is a possibility that further human skeletal remains may be interred elsewhere in the vicinity.

A total of 1828 disarticulated fragments were excavated representing a minimum number of 19 individuals for the collated assemblage or 27 individuals taking discrete contexts and age into account. The age assessment as observed through epiphyseal fusion, long bone length and dental development indicated that 70.4% of the MNI sample was adult (n=19). Sex assessment was restricted to only 4.3% of elements but of the 58 elements that could be assessed either using morphological observation or metric analysis, 72.4% (n=42) were male/possible male. The aging of adults and sub-adults was highly restricted due to the disarticulated and anatomized nature of the assemblage but one old adult, three middle and four young adults were identified along with two neonates, one infant, one young child, two juveniles and one adolescent.[1] A total number of 243 elements exhibited pathological changes that were diagnosed according to congenital, metabolic, inflammatory, neoplastic, trauma or joint disease categorisation. In these elements, over half of the lesions were inflammatory (53.5%), with joint disease (24.3%) and trauma (16.5%) being the next most common types of lesions. The majority of the inflammatory lesions were non-specific but extensive and gross changes were observed in several elements. It is clear that the bones of the lower leg (tibia and fibula) were associated with a very high rate of inflammation (see Table 8.1) reflecting the tendency of the lower leg to be involved in soft tissue trauma and infection.

Table 8.1. True prevalence rates of non-specific inflammatory lesions.

Element	No.	Total No. Of Elements	TPR(%)
Tibia	34	110	30.9
Rib	44	581	7.5
Fibula	13	65	20
Femur	9	114	7.9
Radius	5	54	9.3
Ulna	4	57	7
Humerus	4	77	5.2
Clavicle	2	25	8
Ilium	2	61	3.3

Table 8.2. Quantification of individual fragments exhibiting modification.

	Total	Total %	Modified	% Modif. By element	% Modif. Of all elements
Cranium	82	4.5%	26	31.7%	11.0%
Mandible	6	0.3%	2	33.3%	0.8%
Clavicle	25	1.4%	6	24.0%	2.5%
Scapula	47	2.6%	4	8.5%	1.7%
Humerus	77	4.2%	13	16.9%	5.5%
Radius	54	3.0%	8	14.9%	3.4%
Ulna	57	3.1%	7	12.3%	3.0%
UnID Long Bone Frags.	9	0.5%	2	22.2%	0.8%
Hand	82	4.5%	0	0.0%	0.0%
Vertebrae	221	12.1%	44	19.9%	18.6%
Sternum	10	0.5%	0	0.0%	0.0%
Ribs	582	31.8%	17	2.9%	7.2%
Pelvis	80	4.4%	6	7.5%	2.5%
Sacrum	23	1.3%	2	8.7%	0.8%
Femur	114	6.2%	41	36.0%	17.4%
Tibia	110	6.0%	45	40.9%	19.1%
Fibula	65	3.6%	11	16.9%	4.7%
Patella	6	0.3%	0	0.0%	0.0%
Foot	178	9.7%	2	1.1%	0.8%
Total	**1828**	**100.0%**	**236**	**12.8%**	**100.0%**

In addition to the non-specific inflammation, there were two possible cases of tuberculosis of the hip joint, one a rare case of a fistulating abscess of the greater trochanter (see Ortner, 2003) (Figure 8.2), and five elements exhibiting changes associated with venereal syphilis. Lesions on the visceral surfaces of several rib fragments were also noted, indicating inflammation of the pleura. Joint disease was common in the vertebrae but rare in long bones. Several cases of trauma were identified. Interestingly, six peri-mortem/unhealed fractures of long bones were present that are generally rare in skeletal assemblages. One example was a comminuted fracture of the forearm (ulna and radius) associated with discrete, concentrated deposits of woven bone. Re-association of some of the elements suggested that an infection had been contracted in the forearm and hand. A smaller amount of bone formation had occurred at the fracture site over the unhealed breaks but the arm had subsequently been amputated above the elbow, presumably due to the fracture failing to heal and becoming life-threatening. All six unhealed fracture elements are distal portions and it is likely that these represent cases of trauma requiring emergency amputations.

A total of 236 elements exhibited evidence of modification in the form of incisive defleshing marks, cut marks, chop marks and saw marks that were recorded according to Reichs (1998: 359). Skeletal elements from almost all areas of the body exhibited modification, though the sterna, patellae and hand elements were unaffected (See Table 8.2).

Figure 8.2. Fistulating abscess of the greater trochanter, possibly linked to tuberculosis. Note the false start kerf across the femoral neck [785].

The most frequently modified elements were those of the cranium, vertebrae and the leg bones; more specifically, the tibia was the most frequently modified element, with 40.9% of all tibiae in the assemblage having undergone some form of modification. Of all the femora, 36.0% were modified and of the mandible and cranium, 33.3% and 31.7% had been modified respectively. Of fragments that were identified as sub-adult (N=101), 23.3% exhibited modifications compared to 13.6% of identifiable adult elements (N=1260).

The distribution of cut marks about the body suggests that the assemblage consisted of a high proportion of bisected or amputated long bones and craniotomies with a smaller proportion of anatomized post-cranial elements. Some of the cuts on individual elements were numerous and elaborate, particularly in the spine and the cranium. Three long bone elements had been bisected at both ends of the diaphysis and several axial elements such as the mandible, vertebrae and bones of the pelvic girdle exhibited cuts for anatomical dissection, including transverse bisection at the midline. Other such modifications included cuts across the mandibular and pubic ramus as well as removal of portions of frontal or occipital bone, all of which had clearly been undertaken in order to view underlying soft tissue structures *in situ* (Figure 8.3). Defleshing marks indicated by incisive knife cuts were present on several elements, particularly associated with muscle attachment sites.

By analysing the distribution of elements and modifications according to archaeological context, further insight is gained into the nature of the non-normative deposits at the Infirmary. For example, context (5008) was one of several fills within the deep pit and contained relatively few elements, the majority of which were long bones. Context (5003) formed the complete fill of a large shallow pit and constituted the majority of the assemblage. Although most areas of the body were represented, a high

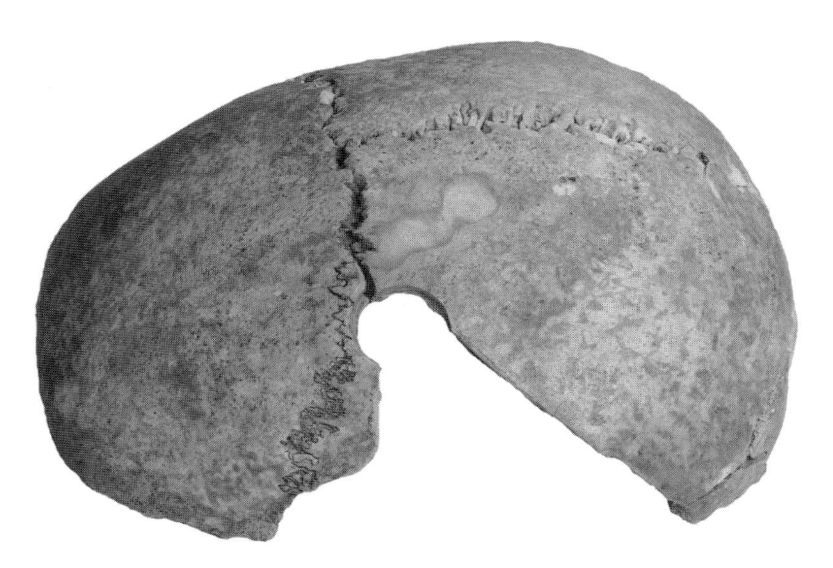

Figure 8.3. Trepanation and viewing aperture cut into the calvarium [1287].

proportion of the elements were long bones, whereas axial and foot elements were relatively under-represented in this context. In comparison, the remains contained in context [10003], the shallow pit located in the front garden, were more representative of all areas of the body. A total of two adult individuals were represented by these remains in addition to one sub-adult, the remains of the latter consisting of an amputated forearm. Closer examination of the adult elements suggested that several may well have been associated with the same adult male individual, although it would not be possible to confirm this without DNA analysis.

Examining the distribution of modified elements, context [5003] contained 88.9% of all bisected femora and 100% of the bisected tibiae, the most commonly modified elements in the assemblage. In contrast, 86.1% of bisected vertebrae and 60% of bisected ribs were found in context [10003]. This pattern in the distribution of elements and modified bones suggests that context [5003] consisted of a large, single deposit of bisected long bones in addition to anatomized disarticulated elements. The scale and nature of the deposit suggests that these remains had been retained for some time and possibly represented a teaching collection of preserved specimens and prosections. Although context [5008] contained comparable skeletal elements and modifications, the scale of the deposit was much smaller and given that it appears to have been discarded as one deposit within a large rubbish pit, it is likely to represent surgical waste along with a small quantity of anatomized elements that had not been retained for any particular length of time, indicating a practice of intermittent clearance and disposal of body parts from the dissection room. In contrast, the range of elements and the higher frequency of dissection cuts suggest that context [10003] contained the partial remains of at least one anatomically dissected adult male, along with additional anatomized adult remains and an amputated arm of a sub-adult.

4. Bioarchaeological Reflections of Anatomized Remains

The bioarchaeological evidence from the skeletal assemblage found in discrete waste pits at the Worcester Royal Infirmary provides intriguing evidence of the prevailing attitudes towards the body and body parts. From 1818, the infirmary was the seat of Charles Hastings who acted as a medical officer for the city as well as physician at the Infirmary. Whilst campaigning for medical care for the poor, he was also an ardent campaigner for ethical medical practice within the profession and to this end founded the British Medical Association at Worcester. It is no co-incidence that the inaugural meeting was held on the very day the Anatomy Act was passed, 19th July 1832. The passing of the Act marked a new dawn in the rise of the recognition of medicine as a professional practice. From this platform the unity and viability of the discipline was to be consolidated by the BMA through promoting ethical conduct, including the legitimate procurement of bodies for dissection. Medical and humanitarian ethics embraced the utilitarian movement of the day; the exchange of aid for dead bodies and organs was justified in view of the contribution this made to wider society, irrespective of the personal wishes of the deceased. However, the reputation of the hospital was prominent in the minds of the practitioners and they went to great lengths to overtly dissociate themselves with death, including deciding against establishing a burial ground in 1814 despite the Infirmary being located in open grounds. The discovery of the skeletal assemblage begs the question as to whether illicit dissection and concealment of human body parts had occurred in order to maintain the reputation of the hospital under the immense pressure to achieve medical excellence.

First and foremost the hospital sought to alleviate the symptoms of admitted patients. Evidence for amputation is abundant. The analysis shows that 79.2% of bisected tibiae and 75.0% of bisected fibulae exhibited macroscopic non-specific inflammatory lesions. These elements may well represent the 'Inflamed Legs' mentioned by Hastings (1841). So great was the incidence of ulcerated legs at the Infirmary that in 1818, two attics were fitted up with ten beds as a septic ward so that these patients could be treated separately (McMenemey, 1959). Many of these bisected and diseased lower limb elements in the skeletal assemblage are likely to represent surgical amputations (Figure 8.4). Amputation was viewed as a 'mutilation of the body' and was only to be undertaken as a last resort for the removal of limbs that were 'worse than useless' (Liston, 1842: 612). Liston (1842) recommended such intervention in life-threatening cases of bad and complicated fractures, diseased joints and diseased bone (such as those affected by malignant tumours and ulcers), chronic suppuration and badly formed or infected stumps. In this context, it could be expected that a high frequency of amputated limbs would exhibit bone lesions from chronic disease, although malignancies only affecting soft tissue would also have required amputation of the limb (Liston, 1842). BABAO-funded digital radiography undertaken on the bisected limb bones indicated that the majority of lesions were likely to be associated with chronic and acute osteomyelitic infections (Western and Bekvalac, in prep., a).

In Worcester, more males than females underwent amputation for the treatment of disease and trauma (Carden, 1864). These were high risk operations, not only threatening the mortality statistics of the hospital but also potentially rendering a surviving individual incapacitated, compromising self and social identity. As the contemporary observer of London life, Smith (1857: 17), remarks 'Commerce maims and mutilates her victims as effectually as war…; and men and lads without arms, or without legs, or without either, who have yet had the misfortune to escape death, are left without limbs…to fight the battle of life at fearful odds'. Amputation rendered males effeminate and incomplete persons, raising issues over the completeness

Figure 8.4. Amputation of infected lower leg [901].

of bodies, substitutability and the substance of body parts in relation to personhood (O'Connor, 2000; Smith, 2013). Nonetheless, a successful amputation could not only save a life but also render a body functional again through the use of modern prosthetics, 'constructing the fragmented body as essentially whole...[as] a means of restoring the integrity of self' (O'Connor, 2000: 121). In the Industrial period, this integrity of self was evaluated in terms of an individual's physical functionality in order to work and be self-sufficient. Motivated by this, amputation techniques were formulated by selecting specific sections of limb bones for retention and removal in order to facilitate post-surgical rehabilitation. Carden, a surgeon at the Worcester Infirmary from 1838 to 1872, developed a specific technique of amputation through the femoral condyles, known as amputation by single flap, in order to promote healing and render the stump fully weight bearing (Carden, 1864; Western and Kausmally, 2011). An example of this 'Carden Amputation' (Wood, 1872) survived amongst the discarded waste (see Western, 2012).

The radiographic analysis also confirmed that 46.1% of the bisected long bone elements exhibited no pathological change. Whereas 67.6% of those pathological bisected long bone elements were distal portions, the non-pathological elements consisted of approximately equal numbers of proximal and distal parts. Could at least some of these elements represent post-mortem modification as part of dissection or surgical training? Unfortunately, in the majority of cases it is not possible to categorically differentiate between ante-mortem and post-mortem bisections, particularly from disarticulated assemblages. Circular cutting of the limb bone is the most likely to indicate a post-mortem procedure, where rotation of the element has occurred to make the cut by three or more separate horizontal sawing actions. This may indicate that the limb was already significantly reduced in size and volume at the time of the bisection, allowing rotation of the element on a flat surface. Other indications may involve large breakaway spurs or notches, unusual directions in sawing, burnishing of cut surfaces (Witkin, 1997), false start kerfs located some distance away from the margin of the bisection or steps in the outline of the cut. Nonetheless, these signatures in the bone are not unequivocal.

A number of cuts observed in the Infirmary assemblage were clearly, however, not therapeutic in origin. As detailed above, these cuts were undoubtedly for dissection purposes and were dictated by anatomical location for viewing underlying soft tissue structures. There can be ambiguity in interpretation of these modifications, however, since 'dissection' undertaken for morbid pathology, post-mortem and anatomical dissection followed the same methodology, albeit in an *ad hoc* fashion. Irrespective of the nature of the original motive of the modifications, however, bodies should have received a decent burial. Prior to the Anatomy Act 1832, only the bodies of felons were exempt from customary funerary rites in this regard, dissection of the body having been meted out as a punitive measure prohibiting a traditional Christian interment. The dissected remains contained in context [10003] may well represent those of executed felons, the partial remains of one of whom appear to have been discarded together, disposed of in plain sight in the front garden. After the Anatomy Act 1832, it was required that all dissected bodies were to receive a decent burial. Disposal of surgical waste in rubbish pits such as those found at Worcester would have been an appropriate means of discarding rotting limbs but this was certainly not deemed a fitting recompense for the infliction of anatomical dissection upon the body. However, the practicalities of dealing with dissected remains were more complex than anticipated by Warburton's Parliamentary Committee drawing up the Act.

A loophole in the Anatomy Act 1832 allowed body parts to be removed and retained without legal repercussions. Prior to the application of formaldehyde, anatomical dissection would frequently involve the removal of areas

Figure 8.5. Illustration of a pathological element removed for retention following a hospital post-mortem examination by Liston. Amputation for this case of 'spina ventosa' was refused by a female relative of the patient, who subsequently died untreated, on the grounds that 'he was better dead than alive with the loss of one of his extremities' (Liston, 1837: 294).

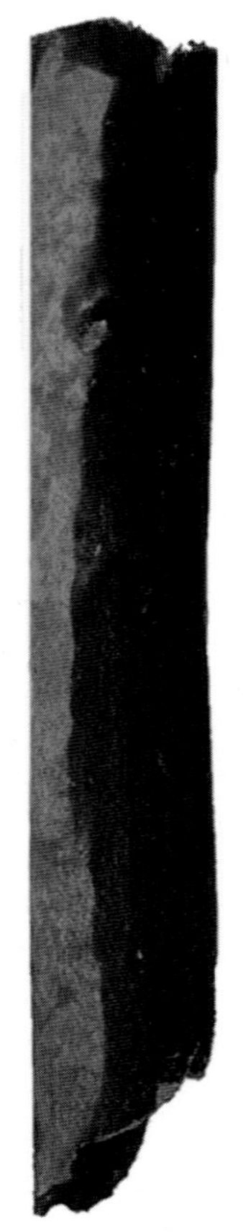

Figure 8.6. Example of a stained element.

of the body that would putrefy the soonest i.e. the brain and abdominal viscera (Parsons, 1831) with some limbs being retained longer as prosections for demonstration purposes. Morbid dissection frequently resulted in body parts being removed for retention in pathology collections for visual inspection and some of these would have consisted of, or at least contained, skeletal elements (Figure 8.5) (Mitchell and Chauhan, 2012). By 1834, gaining the recognition of teaching hospital status required the establishment to hold a collection or museum of anatomical and pathological specimens (Reinarz, 2006) and by 1854, collections at teaching hospitals were regulated and standardised by the Society of Apothecaries and the Royal College of Surgeons (Edwards and Edwards, 1959). Anatomical museums were very much a vehicle for validating the new scientific approach to the body and were often open to the educated public for perusal (Edwards and Edwards, 1959).

Hastings founded an anatomical museum at the Worcester Infirmary in 1813 (McMeneney, 1947), which was still in existence in 1848 (Carden, 1864) though is not extant today. Fifty four elements from the Worcester Royal Infirmary give clear indication of use as teaching specimens. A series of ribs and a femoral head, for example, were recovered with metal fastenings still attached. Some fragments were stained green throughout the cortex and emanated a distinct chemical malodour, some exhibited brown/black staining and other elements were bleached in appearance (Figure 8.6). Amputated limbs were sometimes retained as specimens for the teaching collection at the Infirmary (Carden, 1864) and would have undergone the appropriate preparation and preservation techniques as secondary modifications. However, the origins of teaching specimens and body parts were not necessarily local. The Infirmary committee minutes, for example, document the purchasing of a teaching skeleton from an external source (McMenemey, 1947). Inventories of anatomical specimens were frequently published as catalogues (Edwards and Edwards, 1959) and body parts were regularly traded openly between collectors and institutions (Edwards and Edwards, 1959; Alberti, 2011; Hurren, 2011;Chaplin, 2012; Mitchell and Chauhan, 2012). The trade and collection of body parts by medical practitioners of this era denotes the supplanting of the 'symbolic capital' of these once inalienable possessions by the 'commodity fetishism of capitalism' (Lock, 2001: 69–70), part of the very fabric of Industrial period society. The nouveau medics, whose success in the provinces was particularly dependent upon their own entrepreneurship (Reinarz, 2006), materialised their expertise within the social circles of wealthy literati and patrons via anatomy museum collections (Brown, 2011). As discussed by Alberti (2011), such collections embodied complex networks of social exchange underpinned by the commodification of body parts, while augmenting the status of hospitals, their founders and individual practitioners.

While the preparation and public presentation of body parts in anatomy museums played a crucial role in the naturalisation of the body throughout the Enlightenment (Chaplin, 2012), it was greatly at odds with contemporary cultural attitudes towards the corpse and the afterlife. In the Victorian period, funerals were to be public affairs lavished with monuments and commemorative memorabilia, in the hope of preserving in perpetuity an individual's memory and social status (Laqueur, 1983). These monuments served as a material reminder of the achievements and aspirations of a life well lived, locating the body in its earthly grave and signifying by proxy a metamorphosis into the ascended Aristotelian *anima*. Preserved anatomical specimens, conversely, are the materialisation of perpetual anonymity; indeed, one of the major underpinnings of medical ethics is the maintenance of privacy and confidentiality (Jones *et al.*, 2003). In opposition to *memento mori*, the preservation of human body parts for direct visual appreciation served to create an 'aesthetic of absence' (Ash, 1996 after Hallam and Hockey, 2001: 50) not only of personhood contained by the enveloped body whole but also of death itself. This stasis of the artificially exposed, broken down body challenged the accepted natural progression of corporeal decay and simultaneous spiritual release from earthly sins through death (Hallam and Hockey, 2001). It halted transcendence, transfixed observers and provoked their moral judgement on 'where the body is bounded in time, space, or form' (Landecker, 1999, after Lock 2001: 74). This new debate sparked by medical epistemology promoted an emphasis on the corporeality of existence and personhood (Foucault, 2003), resulting in a growing secular trend towards biological determinism (see Gowland and Thompson, 2013).

Exacerbating 'anonymisation' brought about by the display of body parts, the lack of consent required by the Anatomy Act 1832 from the deceased for anatomical dissection meant that objectified body parts embodied the complete social failure of an individual. Autonomy, an essential agency of social identity in this new era of mercantile self endeavour and empowerment, was wholly compromised by the act of anatomization via conscription. Post-mortem examinations, for which consent was obtained, became increasingly acceptable due to the perceptibly justified objective of elucidating the cause of death and also to the return of the corpse in a seemingly complete state for deposition in hallowed ground (Crossland, 2009). Integrity of the body was traditionally equated to integrity of personhood; the disposal of severed, disarticulated body parts violated this fundamental tenet of social identity, atomised via anatomization of the body. The destructive nature of dismemberment and the collection of specimens that bodies underwent rendered the 'decent burial' of anatomized remains absurd to some contemporaries (Knott, 1985). Inevitably, at least some of the dissected bodies removed for burial from the Infirmary would not be complete.

Body parts such as those recovered from the waste pits at the Worcester Royal Infirmary may have been retained for lengthy periods of time or periodically disposed of, depending upon the esteem they conferred, their value to the curator (often also the pathologist) and how easily replaced they were (Alberti, 2011). The bioarchaeological evidence from Worcester illustrates that many of the anatomized human remains consisted of adult males whose economic, occupational and personal circumstances at the time of death fell foul of the social system in the Industrial Period (Western and Bekvalac, in prep., b). The involuntary posthumous legacy of those deemed socially redundant was highly prized by medical practitioners, irrespective of the stigma of 'anonymisation' they generated by carrying out the experimental and permanent dismemberment of the deceased. The retention and irreverent disposal of such constituent body parts in waste-pits was considered a justified consequence of placing the authority of public humanitarian objectives over private eschatological beliefs.

5. Conclusion

The discovery of a large disarticulated skeletal assemblage in the grounds of the Worcester Royal Infirmary has provided a unique insight into medical practice during the late Georgian and Victorian periods. It is clear from the assemblage that a large number of elements represent amputated diseased limbs discarded after the therapeutic treatment of patients at the hospital with acute and chronic infections. However, a substantial number of elements were also subject to anatomical and morbid dissection, as well as being used for surgical training, and were likely to represent old teaching specimens and prosections. Similar assemblages have been noted at Nottingham (Chapman, 1997), Newcastle (Chamberlain, 2012) and Oxford (Hull, 2003).

As a provincial hospital and the seat of the BMA, Worcester Infirmary stood as a paragon of ethical practice in contrast to the sullied reputation of earlier metropolitan hospitals (see Western, 2012). However, during both life and death, bodies reduced to components obfuscated personhood in the eyes of the Victorian populace. The non-normative discarding of body parts in waste pits reflects the 'functional illiteracy of the disarticulated body' (O'Connor, 2000: 115) in the schema of biological reductionism. Contrary to contemporary notions of the body and funerary practice (including the decent burial of amputated limbs; see Tarlow, 2010), dissection and amputation naturalised human body parts and rendered them disengaged, 'anonymised' specimens. This allowed the medical practitioner to openly retain, reconfigure and publically display involuntarily donated body parts to inspire and promote medical education under the auspices of humanitarianism. A traditional Christian burial for such objectified specimens was not seen as an appropriate nor practical means of disposal. Even to this day, the treatment of anatomical specimens over 100 years old remains outside of legislation, testifying to the unresolved ethical issues regarding the fate of historical 'anonymised' body parts (Jones *et al.*, 2003). This is in some part due to the discussion about rights, ownership and body boundaries, a debate initiated by the inception of human dissection in medicine that continues today focussing on the controversial ramifications of the archiving of genetic human material (Lock, 2001) as well as the repatriation of museum 'specimens' (see, for example, Carvajal, 2013). However, the lacuna in legislation is mainly a result of ethical shifts in medical practice and what are now perceived of as the dubious circumstances in which many of these 'regrettable' anonymous specimens were obtained without consent during post-mortem procedures (Jones *et al.*, 2003: 346).

Acts of post-mortem examinations in the Industrial period primarily served to reincorporate the unidentified or dislocated individual into the social system according to the utilitarian and humanitarian principles prevalent of the period, mediated by economic status, even after death. As observed by Good (2003: 70) 'medicine as a form of activity joins the material to the moral domain'. The identification of pathological conditions, in both medical and forensic cases, required the physical reduction of the individual to visually observable body components in order to achieve social and medico-legal justice. At the same time, society reduced its ills to the level of the individual; usually, though not exclusively, selecting the criminal, the poor, the ill and the unknown, whose bodies manifested the lesions of the hazardous workplace, squalid living conditions and accidental or violent trauma. These alienable bodies filled the dire need for inspection of the corpse by philanthropic medical pioneers. While these selected individuals had little control over their posthumous fate or their own bodies and body parts, in the hands of the anatomists these took on new directions and social significance. This appreciably impacted on how they were consumed and disposed of by Victorian society (Alberti, 2011) and furthermore, how they are subsequently rediscovered, re-evaluated and reincorporated into our archival social body today. Such retention, display and disposal of human bodies and body parts continues to bring to the fore the tensions created between medical ethics, social norms and humanitarian objectives (see, for example, Calain, 2013; Clark, 2013).

Overall, post-medieval infirmary assemblages provide a wealth of data regarding surgical intervention, such as amputation, as well as disease and trauma that are rarely observed in contemporary skeletal populations. Upon the excavation of such assemblages, the archaeological context of the modified human remains is paramount to our understanding of their nature, as osteological signatures for amputation, bisection, post-mortems, anatomical and morbid anatomy are not unequivocal in many cases. It is the combination of bioarchaeological observations for post-mortem intervention and non-normative modes of deposition that gives us a direct insight into evolving attitudes towards the body during the period. These long forgotten severed body parts illuminate the social context of medical science during the Industrial Period and uniquely convey how the visual perspicuity of body parts ultimately allowed medicine to replace the blind faith of religion as the 'core of our soteriological vision' (Good, 2003: 70) in humanitarianism today.

Acknowledgements

Ossafreelance gratefully acknowledges the assistance and co-operation of the University of Worcester in the preparation of this article. Any errors or omissions remain the responsibility of Ossafreelance. The author is also particularly grateful to Tania Kausmally who co-authored the osteological report upon which the work presented here is based and to Mark Farmer who undertook the digital radiographic imaging on behalf of Reveal Imaging Ltd. Special thanks also go to Jelena Bekvalac (Museum of London), Dr. Becky Gowland (Durham University), Dr. Sam Alberti (Royal College of Surgeons, London) and an anonymous reviewer for their comments on an earlier version of this paper, to Derek Hurst and Jane Evans (Worcester Archive and Archaeology Service) for the provision of contextual data and to Cathy Patrick (CgMS Consulting) for facilitating this publication.

Note

1. Age Categories: Foetal = <36 weeks; Neonate = 0–1 month; Infant = 1–12 months; Young Child = 1–3 years; Older Child = 4–7 years; Juvenile = 8–12 years; Adolescent = 13–19 years; Young Adult = 20–34 years; Middle Adult = 35–49 years; Old Adult = 50+ years.

References

Alberti, S.J.M.M. 2011. *Morbid Curiosities: Medical Museums in Nineteenth Century Britain.* Oxford University Press, Oxford.

Bartrip, P.W.J. 2003. *The Home Office and the Dangerous Trades; Regulating Occupational Disease in Victorian and Edwardian Britain.* Editions Rodopi B. V., Amsterdam and New York, NY.

Binski, P. 1996. *Medieval Death.* British Museum Press, London.

Brown, M. 2011. *Performing Medicine: Medical Culture and Identity in Provincial England c. 1760–1850.* Manchester University Press, Manchester.

Burney, I.A. 1994. Viewing bodies: medicine, public order and English inquest practice. *Configurations* 2, 33–46.

Calain, P. 2013. Ethics and images of suffering bodies in humanitarian medicine. *Social Science & Medicine* 98, 278–285.

Carden, H. 1864. On amputation by single flap. *British Medical Journal* 1(172), 416–421.

Carvajal, D. 2013. Museums Confront the Skeletons in Their Closets. *The New York Times* (online). Accessed July 9, 2013. http://www.nytimes.com/2013/05/25/arts/design/museums-move-to-return-human-remains-to-indigenous-peoples.html?pagewanted=1&_r=2.

Chamberlain, A.T. 2012. Morbid osteology: Evidence for autopsies, dissection and surgical training from the Newcastle infirmary burial ground (1753–1845), in: Mitchell, P. (Ed.), *Anatomical Dissection in Enlightenment England and Beyond: Autopsy, Pathology and Display*. Ashgate Publishing, Surrey, pp. 11–22.

Chaplin, S. 2012. Dissection and display in eighteenth-century London, in: Mitchell, P. (Ed.), *Anatomical Dissection in Enlightenment England and Beyond: Autopsy, Pathology and Display*. Ashgate Publishing, Surrey, pp. 95–114.

Chapman, S.J. 1997. The findings of a possible reference collection in the grounds of a Victorian general hospital, Nottingham, UK. *Journal of Paleopathology* 9, 37–46.

Cherryson, A. 2010. In pursuit of knowledge: dissection, post-mortem surgery and retention of body parts in 18th and 19th century Britain, in: Rebay-Salisbury, K., Stig Sorensen, M.L., Hughes, J. (Eds.), *Body Parts and Bodies Whole: Changing Relations and Meanings*. Oxbow Books, Oxford.

Cherryson, A., Crossland, Z., Tarlow, S. 2011. *A Fine and Private Place: The Archaeology of Death and Burial in Post-Medieval Britain and Ireland*. Leicester Archaeology Monograph 22, 4word Ltd, Bristol.

Clark, N. 2013. Dead Serious? Photo of Damien Hirst with Severed Head Riles Richard III Academics. *The Independent* (online). Accessed July 15th, 2013. http://www.independent.co.uk/arts-entertainment/art/news/dead-serious-photo-of-damien-hirst-with-severed-head-riles-richard-iii-academics-8706571.html

Crossland, Z. 2009. Acts of estrangement: The post-mortem making of self and other. *Archaeological Dialogues* 16, 102–125.

Edwards, J.J., Edwards, M.J. 1959. *Medical Museum Technology*. Oxford University Press, London.

Evans, C.J. In prep. Associated Finds from the Worcester Royal Infirmary, Castle Street, Worcester. Worcester Archive and Archaeology Service. WCM101627.

Foucault, M. 2003. *The Birth of the Clinic: An Archaeology of Medical Perception*. Routledge Classics, Abingdon and Oxford.

Fowler, L., Powers, N. 2012. *Doctors, Dissection and Resurrection Men: Excavations in the 19th Century Burial Ground of the London Hospital, 2006*. MOLA Monograph 62, Museum of London Archaeology, London.

Good, B.J. 2003. *Medicine, Rationality and Experience: An Anthropological Perspective*. Cambridge University Press, Cambridge.

Gowland, R., Thompson, T. 2013. *Human Identity and Identification*. Cambridge University Press, Cambridge.

Hallam, E., Hockey, J. 2001. *Death, Memory and Material Culture*. Berg, New York, NY.

Hamlin, C. 1986. Scientific method and expert witnessing: Victorian perspectives on a modern problem. *Social Studies of Science* 16, 485–513.

Hastings, C. 1827. On the peculiarly soft state of the structure of the lungs. *Edinburgh Journal of Medical Science* 5(3), 1–21.

Hastings, C. 1841. Worcester infirmary: a report of cases attended at this hospital. Journal of the *Provincial Medical and Surgical Journal* S1–1/21, 342–344.

Hastings, C.1855. Facts illustrative of cerebral pathology. *Association Medical Journal* 145, 925–928.

Henderson, D., Collard, M., Johnston, D. 1996. Archaeological evidence for 18th-century medical practice in the Old Town of Edinburgh: Excavations at 13, Infirmary Street and Surgeon's Square. *Proceedings of the Society of Antiquaries Scotland* 126, 929–941.

Holden, L. 1851. *A Manual of the Dissection of the Human Body*. Highley and Son, London.

Hull, G. 2003. The excavation and analysis of an 18th-century deposit of anatomical remains and chemical apparatus from the rear of the first Ashmolean Museum (now the Museum of History of Science), Broad Street, Oxford. *Post-Medieval Archaeology* 37, 1–28.

Hurren, E. 2004. A pauper dead-house: the expansion of the Cambridge anatomical teaching school under the late Victorian Poor Law, 1870–1914. *Medical History* 48, 69–94.

Hurren, E. 2009. Remaking the medico-legal scene: A social history of the late Victorian coroner in Oxford. *Journal of the History of Medicine and Allied Science* 65, 207–252.

Hurren, E. 2011. *Dying for Victorian Medicine: English Anatomy and its Trade in the Dead Poor, c. 1834–1929*. Palgrave Macmillan, Basingstoke, Hampshire.

Jones, D.G., Gear, R, Galvin, K.A. 2003. Stored human tissue: An ethical perspective on the fate of anonymous, archival material. *Journal of Medical Ethics* 29, 343–347.

Knott, J. 1985. Popular attitudes to death and dissection in early nineteenth century Britain: The Anatomy Act and the poor. *Labour History* 49, 1–18.

Lane, J. 2001. *A Social History of Medicine: Health, Healing and Disease in England 1750–1950*. Routledge, London.

Laqueur, T. 1983. Bodies, death and pauper funerals. Representations 1, 109–131.

Laqueur, T. 1989. Bodies, details, and the humanitarian narrative, in: Hunt. L. (Ed.), The New Cultural History. University of California Press, Berkeley and Los Angeles, pp. 176–204.

Liston, R. 1837. *Practical Surgery*. J. Churchill, London.

Liston, R. 1842. *Elements of Surgery*, 2nd edition. Barrington, Ed. and Haswell, Geo. D., Philadelphia, PA.

Lock, M. 2001. The alienation of body tissue and the biopolitics of immortalized cell lines. *Body and Society* 7(2–3), 63–91.

Manuel, D.E. 1996. *Marshall Hall (1790–1857): Science and Medicine in Early Victorian Society*. Rodopi, Amsterdam and Atlanta, GA.

McMenemey, M. 1947. *A History of the Worcester Royal Infirmary*. Press Alliances Ltd., Fleet Street, London.

McMenemey, M. 1959. *The Life and Times of Sir Charles Hastings, Founder of the British Medical Association*. Livingstone, Edinburgh.

Mitchell, P.D., Chauhan, V. 2012. Understanding the contents of the Westminster hospital pathology museum in the 1800's, in: Mitchell, P. (Ed.), *Anatomical Dissection in Enlightenment England and Beyond: Autopsy, Pathology and Display*. Ashgate Publishing, Surrey, pp. 139–154.

O'Connor, E. 2000. *Raw Material: Producing Pathology in*

Victorian Culture. Duke University Press, Durham, NC and London.

Ogden, A., 2003 *Skeletal Report for the Tallow Hill Excavation*. BARC Unpublished Report, Dept. of Archaeological Sciences, University of Bradford.

Ortner, D.J. 2003. *Identification of Pathological Conditions in Human Skeletal Remains*. Academic Press, Smithsonian Institution, Washington, DC.

Parsons, U. 1831. *Directions for Making Anatomical Preparations; Formed on the Basis of Pole, Marjolin and Breschet, and Including the New Method of Mr. Swan*. Reproduced 2008, King's Press, Dayton.

Peacock, T.B. 1860. Case of bronzed skin – death – disease of both supra-renal capsules. *Medical Times and Gazette* 1, May, 446.

Reichs, K.J. 1998. Postmortem dismemberment: Recovery, analysis and interpretation, in: Reichs, K. (Ed.), *Forensic Osteology: Advances in the Identification of Human Remains*, 2nd edition. Charles C. Thomas, Springfield, IL, pp. 353–388.

Reinarz, R. 2006. Towards a history of medical education in provincial England. *Medical Historian: The Bulletin of the Liverpool Medical History Society* 17, 30–37.

Reiser, S.J. 1978. *Medicine and the Reign of Technology*. Cambridge University Press, Cambridge.

Richardson, R. 1988. *Death, Dissection and the Destitute*. Phoenix Press, London.

Risse, G.B. 2010. *Hospital Life in Enlightenment Scotland: Care and Teaching at the Royal Infirmary of Edinburgh*. Createspace, North Charleston, SC.

Smith, C.M. 1857. *Curiosities of London Life or Phases, Physiological and Social, of the Great Metropolis*. W. and F. G. Cash, London.

Smith, C.M. 2013. An Essay on Wooden Legs, with Some Account of Herr Von Holtzbein [Excerpt], in: *Nineteenth-Century Disability: A Digital Reader*. Accessed July 2, 2013. http://www.nineteenthcunturydisability.org/items/show/24.

Strange, J.M. 2005. *Death, Grief and Poverty in Britain 1870–1914*. Cambridge University Press, Cambridge.

Tarlow, S. 2010. *Ritual, Belief and the Dead in Early Modern Britain and Ireland*. Cambridge University Press, Cambridge.

Virchow, R. 1880. *A Description and Explanation of the Method of Performing Post-Mortem Examinations in the Dead-House of the Berlin Charité Hospital, with Especial Reference to Medico-Legal Practice, from the Charité-Annalen*. Translated by T. P. Smith. Churchill, Oxford.

Western, A.G. 2006. *Osteological Analysis of the Human Remains from St. Andrew's Burial Ground, Worcester*. Ossafreelance Unpublished Report OGW1011.

Western, A.G. 2012. A Star of the First Magnitude: Osteological and historical evidence for the challenge of provincial medicine at the Worcester royal infirmary in the nineteenth century, in: Mitchell, P. (Ed.), *Anatomical Dissection in Enlightenment England and Beyond: Autopsy, Pathology and Display*. Ashgate Publishing, Surrey, pp. 23–42.

Western, A.G., Bekvalac, J. In prep. a. Historical and Medical Contextualisation of Surgical Waste from the 19th Century Worcester Royal Infirmary using Digital Radiography. (BABAO funded).

Western, A G., Bekvalac, J. In prep. b. The Making of Man: Engendering the Post-Mortem Fate of Bodies in Industrial Period England.

Western, A.G., Kausmally, T. 2011. *Osteological Analysis of the Human Remains from the Worcester Royal Infirmary, Castle Street, Worcester*. Ossafreelance Unpublished Report OA1030.

Witkin, A. 1997. *The Cutting Edge: Aspects of Amputations in the Late 18th and Early 19th Century*. Unpublished MSc Dissertation, University of Sheffield, Sheffield.

Wood, E. 1872. Amputation at the knee joint. *Boston Medical and Surgical Journal* 86, 1–5.

9. Thomas Henry Huxley (AD 1825–1895): Pioneer of Forensic Anthropology

Stefanie Vincent and Simon Mays

Thomas Henry Huxley was a leading Victorian biologist. He is best remembered today for his vocal support for Darwin's ideas on evolution, but we analyse archival documentation which shows that he was also an early pioneer of the forensic application of anatomical and anthropological techniques. Recent work on the skeletal remains of a member of Sir John Franklin's disastrous 1845 expedition to the Arctic led to the discovery of work conducted by Huxley on the skeleton in 1872, shortly after its recovery. Huxley's report on the remains shows he approached the analysis in a manner easily recognisable to modern forensic anthropologists. This is one of the earliest known attempts to use anatomical techniques to formally identify skeletonised remains. Huxley's information was used by the Admiralty as the basis for their identification of the remains as Henry Le Vesconte, a lieutenant on the expedition. Recent re-examination of the remains supports Huxley's principal observations and his report remains an impressive early example of forensic anthropology. However stable isotope analysis shows that the Admiralty's identification, made in the light of Huxley's report, is unlikely to be correct.

Keywords: Forensic Anthropology; Identification; Human remains; Sir John Franklin

1. Introduction

Thomas Henry Huxley (Figure 9.1) was a leading Victorian biologist, best known for his vocal support of Darwin's theory of evolution and as a proponent of universal education (Bibby, 1972; Desmond, 1997). He was born in 1825 in Ealing, west London, and after being apprenticed to a doctor at the age of twelve went on to formal medical training at Charing Cross Hospital in 1842 (Chalmers-Mitchell, 1900). After completing his medical training he enlisted in the Royal Navy and obtained his first post as assistant surgeon on H.M.S. *Rattlesnake,* which was on a mission to survey the coast of Australia. Huxley's naturalist studies on board included observations of microfauna and jellyfish, and led to the publication of his first paper, '*On the Anatomy and the Affinities of the Family of Medusae*,' by the Royal Society in 1849 whilst the voyage was still underway. When he returned to England the Navy granted him a paid leave of absence, which allowed him to publish his seminal work, '*The Oceanic Hydrozoa*' in 1859 (Bibby, 1960). The following year was that of the famous 'Oxford Debate' in which Huxley defended Darwin's ideas on evolution from attack by the Bishop of Oxford. From 1854 Huxley held several academic posts but continued to be linked throughout his career to the Royal Navy. He expanded his early work in comparative anatomy to vertebrates including humans, working within the scientific community to encourage the teaching of practical anatomy to science students. This led to an interest in ancient human remains and in 1863 he published, 'Evidence as to Man's Place in Nature', which contained discussion of European fossil hominin remains, including the Neanderthal cranium from Feldhofer Cave. He also conducted osteological studies on skeletons recovered from archaeological sites, most notably prehistoric remains from Caithness, Scotland (Laing and Huxley, 1866). Furthermore, Sir Arthur Keith attributed the origination of the Frankfort Plane in craniometry not to von Ihering in 1872, but to Huxley eight

Figure 9.1. Portrait of Thomas Henry Huxley, reproduced with permission of the National Portrait Gallery, London.

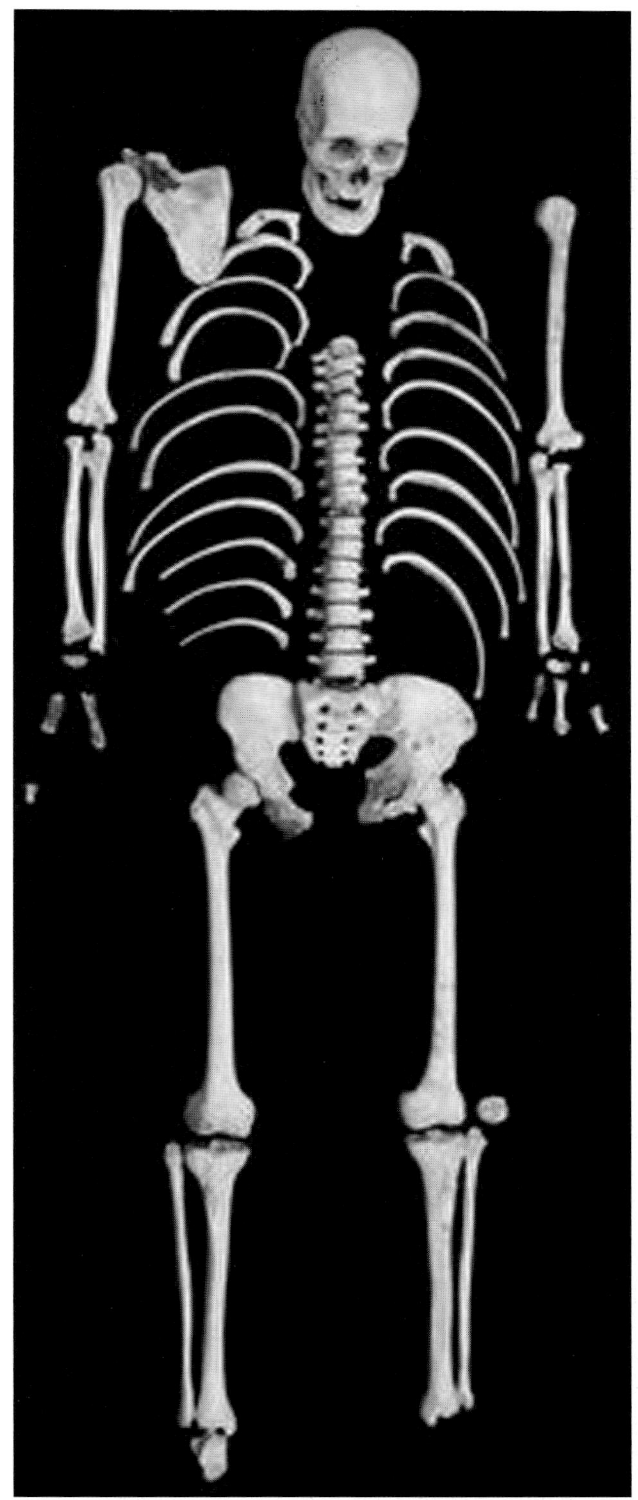

Figure 9.2. Skeleton recovered from King William Island.

years earlier (Bibby, 1972). In 1872 Huxley was teaching classes at the Royal School of Mines when the Admiralty called upon him to help them to identify the skeleton of an unknown sailor recovered from the Canadian arctic.

2. The Skeleton

The skeleton was discovered on King William Island in the Canadian High Arctic, in 1869 by an American adventurer called Charles Francis Hall. Hall's was the latest in a string of expeditions attempting to discover the fate of Sir John Franklin's voyage in search of the North-West Passage, a sea route connecting the North Atlantic with the North Pacific Ocean. The Franklin expedition left England in May 1845, with two ships, the H.M.S *Terror* and H.M.S. *Erebus,* and 129 men. When the expedition failed to return by 1848 the first of several search parties was despatched. Franklin and his companions were officially declared deceased in service by the Royal Navy in 1854, but such was the importance attached to the expedition and the public interest in discovering its fate, that over the following decades several attempts were made to discover the route taken by the expedition and the reasons for its failure. Successive searches found evidence that the crew deserted their ships after they became trapped in ice and had attempted to walk overland to reach help (Cyriax, 1939; Owen, 1978; Lambert, 2009). By the time of Hall's 1869 expedition there was no realistic chance of locating survivors (although Hall appears to have held this hope at the beginning of the trip), however his findings added a new level of detail to the attempts made by the last surviving crew members to reach safety. Hall was led to a grave by local Inuit and discovered an almost complete skeleton (Figure 9.2), which he recovered and which was, on his return, deposited with the Naval Attaché to Washington, Admiral Inglefield (Woodman, 1991).

The skeleton was delivered to Admiral George Henry Richards, Naval Hydrographer at the Admiralty in London, who at first suggested immediate burial would be the best course of action, "...for it could not certainly be identified and would be no use to anyone" (Richards, 29th April 1872). However public interest in the fate of the expedition was still high and after the delivery of the skeleton to his office Richards found himself besieged by interested members of the Admiralty (Richards, 29th June 1872). Soon after, Richards approached Huxley about the matter writing, "I have thought that perhaps you might feel sufficiently interested in the matter to put him together with a view to identification...considering that there are not above half a dozen officers of which to select from I think identification certain" (Richards, 22nd June 1872).

Items of clothing found with the skeleton left the Admiralty in no doubt that it was that of an officer of Franklin's expedition (Woodman, 1991). Remnants of a uniform matched those of Franklin's men and in a letter to Lady Jane Franklin, Richards identified, "...the remains of a silk under vest...", as another reason for their identification of the skeleton as a member of the officer class (Richards, 29th April 1872). The skeleton was delivered to Huxley at the end of June 1872 and the following month he returned a report on it to the Admiralty.

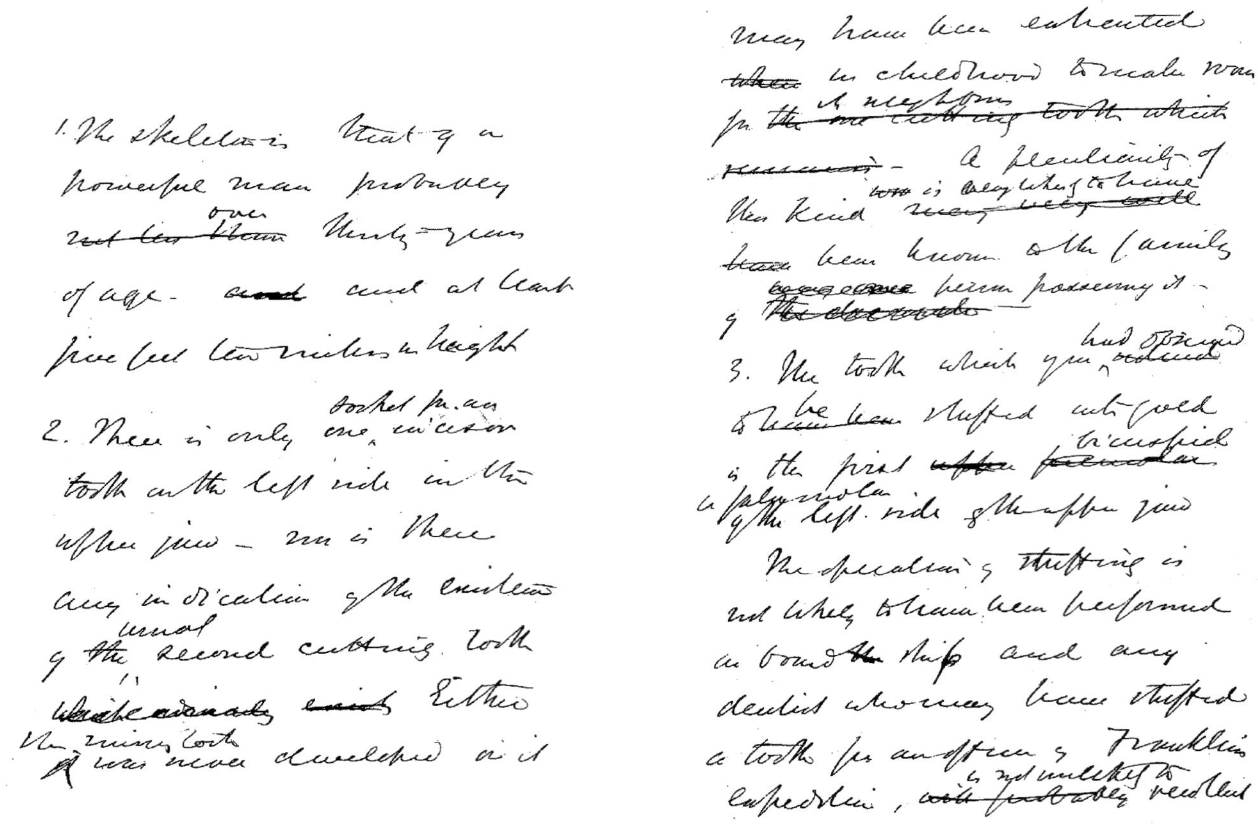
1. The skeleton is that of a powerful man probably over thirty years of age and at least five feet ten inches in height

2. There is only one socket for an incisor tooth on the left side in the upper jaw – nor is there any indication of the existence of the usual second cutting tooth. Either the missing tooth was never developed or it may have been extracted in childhood to make room for its neighbours – A peculiarity of this kind is likely to have been known to the family of person possessing it –

3. The tooth which you had observed to be stuffed with gold is the first bicuspid or false molar of the left side of the upper jaw

The operation of stuffing is not likely to have been performed on board ship and any dentist who may have stuffed a tooth for an officer of Franklin's expedition, is not unlikely to recollect

Figure 9.3. Huxley's handwritten report to the Admiralty, reproduced with permission of the College Archives, Imperial College, London.

3. The Report

We have located both Huxley's own notes on the skeleton (Huxley, 1872a) and the official letter which communicated his findings to the Admiralty (Huxley, 1872b; Figure 9.3). In his own notes, Huxley sets out a detailed account of his findings, while using the official report to distil such information which he felt could be used for identification purposes.

Huxley begins his notes by taking a brief inventory of the remains, detailing those which are connected by (or have adhering to them) desiccated soft tissue, those bones which have been broken post-mortem or are missing, and the general condition of the bones.

Huxley concentrates his detailed examination on aspects which he felt could be of help in identifying the skeleton. He assessed the form of the skull, in order to identify physical attributes and he paid particular attention to the dentition. Huxley's examination of the dental arcade notes that, 'There is a socket for only one cutting tooth on the left side of the upper jaw, and that a high quality gold filling in the first upper right premolar, …leads to the conclusion that this person may have been an officer'.

Whilst Huxley uses his notes to form his thoughts he presents his final report to the Admiralty in a series of succinct points (Figure 9.3), concentrated on aspects which he felt would aid in identification:

"1. The skeleton is that of a powerful man probably over thirty years of age and at least five feet ten inches in height.

2. There is only one socket for an incisor tooth on the left side of the upper jaw – nor is there any indication of the existence of the usual second tooth. Either the missing tooth was never developed or it may have been extracted in childhood to make room for its neighbours. A peculiarity of this kind is likely to have been known to the family of person [*sic*] possessing it.

3. The tooth which you had observed to be stuffed with gold is the first bicuspid or false molar on the left [*sic*] side of the upper jaw.

[4.] The operation of stuffing is not likely to have been performed on board ship and any dentist who may have stuffed a tooth for an officer of Franklin's expedition is not unlikely to recollect the circumstances.

Table 9.1. Comparison of results from Huxley (1872a, b) and Mays et al., *2011.*

	Huxley's Findings	Findings of Mays *et al.*, 2011
Age	30+	30–40
Stature	5′10″	5′8½″ (±1½″)
Dentition	Missing upper left I1, reason uncertain	Agenesis of left maxillary I1 and both mandibular M3s
Pathology	Gold 'stuffed' upper left PM1	Gold filling in upper right PM1, associated file marks on PM1 and adjacent canine.
	Molar in left side slightly decayed.	Large caries cavity in left maxillary M1
		Un-united fracture of the spinous process of T7
Other	Prominent nose and chin Square, powerful jaw	Analysis of cranium classifies individual as white, European

5. The……bones of the face indicate a well formed and prominent nose and chin, and a square set and powerful jaw."

4. The Identification

Following the submission of Huxley's report, the Admiralty arrived at their identification of the skeleton as that of Henry Le Vesconte, a lieutentant aboard one of Franklin's ships. Quite how they arrived at this conclusion is not clear, as papers dealing with the matter have not to date been located. In correspondence Richards offers Huxley his copy of McClintocks, 'Voyage of the Fox', containing details of the members of the expedition (Richards, 24th June 1872) and daguerreotypes had been taken of many of the officers prior to the commencement of the expedition. It appears that Le Vesconte's family were not wholly convinced by the identification[1] and, possibly because of this, when the skeleton was interred under a monument to the lost expedition in Greenwich Royal Naval Hospital, London in 1873, the inscription simply referred to it as of 'one of Franklin's companions'.

5. Recent Analysis

In 2009 the skeleton was made available for examination by osteologists for the first time since 1872. The new analysis and report (Mays *et al.*, 2011), although very different in presentation and precision, highlighted many of the same gross osteological features as Huxley. Table 9.1 compares Huxley's original findings with those of the recent work.

Huxley's report contains no notes regarding his methodology for determination of age at death and stature, so it is impossible to judge the validity of his techniques, however the application of current methods produced results broadly in agreement with his. It is unclear whether Huxley failed to observe some of the pathological aspects present on the skeleton (such as the fracture of the tip of the spinous process at T7) or merely that he did not record them because he felt them of little value in aiding identification. It was only with the aid of forensic facial reconstruction and stable isotope analysis, techniques unavailable to Huxley, that it emerged that the identification as Le Vesconte was unlikely to be correct (Mays *et al.*, 2011). Records show that Le Vesconte was born and brought up in Devon, however stable isotope analysis shows the individual under examination is unlikely to have spent his childhood on the south coast of England. Further to this the forensic facial reconstruction offers a better match to the daguerreotype of HDS Goodsir, a fellow officer aboard the H.M.S. *Erebus*. For a full discussion of these findings, please see Mays *et al.* (2011).

6. Forensic Anthropology in the 19th century

By the time Huxley was approached to study the King William Island skeleton, the practice of applying science to forensic cases had already begun to take hold in the UK. The first UK chair of Forensic Medicine was created at the University of Edinburgh in 1807, swiftly followed by several others (Knupfer, 2000). However, this did not lead to the foundation of forensic anthropology as a discipline in Britain. Forensic anthropology, as practiced today, has its roots in the USA (Brickley and Ferllini, 2007), and during Huxley's lifetime some important developments were taking place there. In 1849, a Harvard Professor, Dr George Parkman, had purportedly been killed by another professor, Dr John Webster. Two of Harvard's anatomy professors examined skeletal remains found at the scene and testified that they belonged to a man aged about 50–60 years, with a height of about 5ft 10 ½" (Burns, 2007: 3; Byers, 2008: 5), consistent with them being Parkman's. A further landmark in the development of the discipline was Dwight's essay on the identification of human skeletal remains in medico-legal contexts, published in 1878 in the USA (Dwight, 1878). He emphasised the importance of a systematic approach, and itemised sex, age and height and the presence of disease or injury or other distinguishing features as key to personal identification from the skeleton. Dwight's publication appears to be an attempt to systematise approaches that were beginning to be used by the nascent discipline in the United States. Huxley's approach to the question of personal identification of the King William Island remains demonstrates the beginnings of similar ways of working in Britain. However, because Huxley's work was unpublished, and his research interests primarily lay in other areas, his contribution went unnoticed by other pioneers of forensic anthropology, and his lead was not followed up in Britain.

The application of forensic anthropological techniques in a high-profile context in the British judicial system would not come until the 1930s when scientific study of dismembered remains found dumped in a ravine in Scotland helped secure the conviction of Dr Buck Ruxton for a double murder (Maples, 1994: 94–98; Cox, 2009: 31).

7. Conclusion

Huxley's work on the skeleton recovered by Charles Francis Hall from King William Island allowed him to provide the Royal Navy with a list of osteological observations upon which to base their identification of the individual. Whilst recent re-analysis has proven that the Admiralty's identification of the remains as Henry Le Vesconte is unlikely to be correct, this does not reflect problems with Huxley's osteological study, which in essence remains sound.

Huxley's work on the King William Island skeleton is an important early application of forensic anthropology, and his brief involvement in this field is an aspect of his work generally overlooked by biographers (e.g. Huxley, 1900; Bibby, 1972). It also shows that the concept of forensic identification was being accepted by governmental authorities in Britain before it came to the consciousness of the general public, and this case appears to be the earliest use of forensic osteology in Britain.

The details of Huxley's contribution to forensic anthropology is a little known episode in the historical development of the discipline. It also provides a useful insight into a scientist whose other accomplishments have served to overshadow his anthropological work.

Acknowledgements

The authors would like to thank William Battersby, Ann Savours and David Woodman for their advice. The help provided by archivists at the National Archives – Kew, Imperial Collage London, the Royal Geographic Society and the Scott Polar Research Centre Cambridge was also invaluable.

Note

1. Details from an unpublished letter in possession of the Le Vesconte family; written to Rose Le Vesconte by E. P. LeFeuvre 1st May 1873.

References

Bibby, C. 1960. *T.H. Huxley: Scientist, Humanist and Educator*. Horizon Press, New York.

Bibby, C. 1972. *Scientist Extraordinary: The Life and Work of Thomas Henry Huxley 1825–1895*. Pergamon, Oxford.

Brickley, M., Ferllini, R. 2007. Forensic anthropology: developments in two continents, in: Brickley, M., Ferllini, R. (Eds.), *Forensic Anthropology: Case Studies From Europe*. Charles C Thomas, Springfield, IL, pp. 3–18.

Burns, K.R. 2007. *Forensic Anthropology Training Manual*, 2nd edition. Pearson, New Jersey.

Byers, S.N. 2008. *Introduction to Forensic Anthropology*, 3rd edition. Pearson, London.

Chalmers-Mitchell, P. 1900. *Leaders in Science: Thomas Henry Huxley, a sketch of his life and work*. The Knickerbocker Press, New York.

Cox, M. 2009. Forensic anthropology and archaeology: past and present – a United Kingdom perspective, in: Blau, S. Ubelaker, D.H. (Eds.), *Handbook of Forensic Anthropology and Archaeology*. Left Coast Press, Walnut Creek, CA, pp. 29–41.

Cyriax, R.J. 1939. *Sir John Franklin's Last Arctic Expedition. A Chapter in the History of the Royal Navy*. Methuen, London.

Desmond, A. 1997. *Huxley: From Devil's Disciple to Evolution's High Priest*. Addison-Wesley, Reading, MA.

Dwight, T. 1878. *The Identification of the Human Skeleton. A Medico-Legal Study*. David Clapp and Son, Boston, MA.

Huxley, L. 1900. *Life and letters of Thomas Henry Huxley by his son, Leonard Huxley*. Macmillan, London.

Huxley, T. *Notes on a skeleton*. July 1872a. Huxley Collection. Reports. Acc. #33.70 Box Number 33 Series 4. Imperial College London.

Huxley, T. *Letter to Admiral Inglefield*. July 1872b. Sent from London. Huxley Collection. General Correspondence. Acc. #19.5 Box Number 19 Series 1i. Imperial College London.

Knupfer, G.C. 2000. History (a) crime scene sciences, in: Siegel, J.A. Saulko, P.J., Knupfer, G.C. (Eds.), *Encyclopedia of Forensic Sciences*. Academic Press, London.

Laing, S., Huxley, T.H. 1866. *Pre-historic Remains of Caithness; with Notes on the Human Remains by Thomas H. Huxley*. Williams and Norgate, London.

Lambert, A. 2009. *Franklin, Tragic Hero of Polar Exploration*. Faber and Faber, London.

Maples, W.R. 1994. *Dead Men Do Tell Tales*. Doubleday, New York, NY.

Mays, S., Ogden, A., Montgomery, J., Vincent, S., Battersby, W., Taylor. M. 2011. New light on the personal identification of a skeleton of a member of Sir John Franklin's last expedition to the arctic, 1845. *Journal of Archaeological Science* 38, 1571–1582.

Owen, R. 1978. *The Fate of Franklin*. Hutchinson, London.

Richards, Admiral Henry. *Letter to Lady Jane Franklin*. 29th April, 1872. Sent from the Admiralty, London. Collection: GB 15 Sir George Richards. Acc. #248/462/34. Scott Polar Research Institute, Cambridge.

Richards, Admiral Henry. *Letter to TH Huxley*. 22nd June, 1872. Sent from the Admiralty, London. Huxley Collection. General Correspondence. Acc. #25.66 Box Number 25 Series 1r. Imperial College London.

Richards, Admiral Henry. *Letter to TH Huxley*. 24th June, 1872. Sent from the Admiralty, London. Huxley Collection. General Correspondence. Acc. #25.6 Box Number 25 Series 1r. Imperial College London.

Richards, Admiral Henry. *Letter to Sophie Cracroft*. 29th June, 1872. Sent from the Admiralty, London. Collection: GB 15 Sir George Richards. Acc. #248/462/12. Scott Polar Research Institute, Cambridge.

Woodman, D.C. 1991. *Unravelling the Franklin Mystery: Inuit Testimony*. McGill-Queens University Press, Montreal.

10. The Concept of Perimortem in Forensic Science

Douglas H. Ubelaker

The term "perimortem" generally refers to an event occurring at or around the time of death when used in forensic contexts. Skeletal evidence of such events lacks indications of antemortem remodeling, as well as characteristics known to be associated with post-mortem conditions. While such evidence cannot be definitively linked to the death event, it can contribute to the overall interpretation of cause and manner of death. While use of the term perimortem extends back at least into the 1970s, varied interpretations and usage contexts call for clear definitions and discussions to ensure effective communication. In forensic anthropology, the term relates mostly to interpretations of the timing of the event, especially in relation to the condition of the bone at the time alterations were produced.

Keywords: Perimortem; Bone; Trauma; Forensic Anthropology

1. Introduction

In the analysis of skeletonized human remains, alterations may be located that appear to have been produced while those remains were relatively fresh. With such evidence, it usually is not possible to determine if the events producing the alterations were sustained shortly before death, shortly after death or were responsible for death. In the last few decades, forensic anthropologists and others in forensic science have used the term perimortem to refer to the broad period of time that includes these various possibilities. Since forensic anthropologists generally are not responsible for the determination of cause and manner of death, they use this terminology to reflect their interpretations of the timing of the injury. In many cases, such information contributes to the determination of cause and manner of death by forensic pathologists, coroners or others with that responsibility.

Perimortem alterations need to be distinguished from those produced antemortem and postmortem (Ubelaker, 1991; Berryman and Haun, 1996; Galloway *et al.*, 1999; Haglund and Sorg, 2002; Symes *et al.*, 2008, 2012; Loe, 2009; Moraitis *et al.*, 2009; Dutelle, 2011). Antemortem conditions are recognized in skeletal cases by evidence of living bone response. Such a response usually takes the form of evidence of bone remodeling, the associated formation of periosteal new bone or related types of bone alteration that are possible only when the bone consists of living tissue.

Evidence that alterations were produced postmortem consists of coloration contrasts between the altered and original bone surfaces, as well as types of alterations specifically associated with postmortem factors. Examples of the latter include evidence of postmortem chewing by rodents and other mammals, shovel marks sustained during recovery, and alterations related to sun exposure or plant root growth.

Although the term perimortem has proven useful in forensic anthropology, it is less popular in the practice of forensic pathology and may have a different connotation (Passalacqua and Fenton, 2012). For example, Fisher (1973) and the 1999 text "Guide to Forensic Pathology" by Dix and Calaluce provide discussions on antemortem and postmortem alterations, but do not introduce the concept of perimortem. A forensic anthropologist may examine a skeleton and conclude that alterations found are "perimortem." A forensic pathologist may examine the same skeleton, but in consideration of other factors conclude that the alterations were sustained while the person was still alive and were responsible for death. A medical specialist may refer to lesions as being perimortem, if they show signs of occurring during life but do not clearly relate to death.

Recent perspective on the perimortem concept is available from documents generated from the Scientific Working Group for Forensic Anthropology (SWGANTH). In a section labeled "Timing of Injuries" the document notes that "within the anthropological realm, perimortem is determined on the basis of evidence of the biomechanical characteristics of fresh bone and does not take into consideration the death event. Perimortem trauma can be considered a default category in which remains lack evidence of healing and there is no diagnostic taphonomic evidence of postmortem damage" (2011: 4). The document also stresses the importance of providing detailed explanation of the use of the term and the relevant evidence.

2. History of the Use of the Term "Perimortem"

Roots of the general use of the term perimortem extend back at least into the 1970s. Van Hoof (1979) used the term in reference to variables involved in biochemical and physical properties of turkey breast muscle related to the food industry. In the medical literature, La Follette *et al.* (1980) refer to "perimortem diagnosis" in the context of discussion of aortocoronary bypass surgery in elderly patients. In the context of toxicology, Fransioli *et al.* (1980) refer to an initial perimortem study in their discussion of methadone and propoxyphene in stored tissue. Within the field of paleontology, the term appears in reference to factors leading to loss of elements in Merrill and Powell's 1980 study of juvenile conodonts. McManus *et al.* (1985) distinguished perimortem burn wound infections from those listed as the principal cause of death.

In 1983, Turner used the term in a bioarchaeological context to differentiate postmortem alterations made soon after death from those sustained long after death. Additional detail is provided by Turner and Turner (1990) in their study of human remains recovered from Wupatki National Monument in Northern Arizona. Specifically, they used the term to describe cut marks, evidence of burning and smashed bone in associated with a feature referred to as Tragedy House. Within early forensic anthropology literature, the term was conspicuously absent in the classic texts of Krogman (1962) and Stewart (1979), although these books presented minimal discussion of evidence of foul play.

3. Coloration

Patterns of coloration play a key role in differentiating perimortem alterations from those sustained postmortem (Adams, 2007). Throughout the postmortem period, skeletons exposed to the elements sustain progressive staining and color alteration over their exposed surfaces. If a bone is broken late in the postmortem period and immediately prior to recovery, it likely will exhibit a color contrast between the exposed internal bone surfaces at the site of the break and the unbroken external bone surfaces. The break reveals the natural cream-colored internal bone that likely will contrast with the taphonomically altered external surfaces. Such color contrasts are easily recognizable, and enable diagnosis of postmortem alterations.

More challenging are bones broken during the post-mortem period but long before recovery. In such cases, even though the bone may have been broken long after death, the broken surfaces may not display diagnostic coloration contrasts with the unbroken surfaces if after being broken the bones continued to be exposed to taphonomic factors. Bones exposed to the elements for extended periods of time prior to recovery will reveal uniform coloration on both the broken and unbroken bone surfaces. Such uniformity suggests that the bone was broken long before recovery, with the internal broken surfaces exposed to taphonomic factors in a manner similar to unbroken external surfaces.

4. Specific Attributes of Perimortem Fractures

In addition to patterns of coloration, investigators have called attention to specific attributes that are associated with fractures of fresh bone. Examples are provided by Moraitis and Spiliopoulou's 2006 study of contrasting patterns of breakage in fresh vs. dry bone. They note that in fresh bone, the fracture angles tend to be obtuse or acute in contrast to the right angles found in dry bone. In fresh bone, the fracture edge tends to be smooth, sharp and beveled with occasional flaking. In contrast, the fracture edge in dry bone tends to be irregular, jagged and blunt. Unfortunately, histological study by Pechniková *et al.* (2011) revealed that observations at the level of osteons could not distinguish between perimortem and postmortem bone breaks.

In a discussion of perimortem trauma SWGANTH (2011) suggests that diagnostic features also include the plastic response of fresh bone fracture and detection of particular mechanisms such as blunt force, sharp force, projectile, or thermal events. The plastic response refers to the ability of the bone to bend or curl when it retains high moisture content.

5. Survivability of Attributes

As noted above, perimortem thermal events can induce fractures and related alterations in bone. Postmortem thermal events also have potential to disguise perimortem evidence and complicate interpretation. The impact of postmortem thermal events is particularly great on coloration patterns since the uniform staining associated with such events likely will mask pre-existing diagnostic coloration patterns.

Considerable research indicates, however, that at least some evidence of perimortem trauma can survive postmortem thermal events and enable detection. Examples originate from both experimentation and case work (Herrmann and Bennett, 1998; Pope and O'Brian, 2004; Symes *et al.* 2005a, 2005b).

6. Early Antemortem Bone Response

As noted above, the distinction between the diagnosis of perimortem and antemortem bone alterations consists of evidence of antemortem bone remodeling or related response. Obvious examples of antemortem extensively remodeled fractures are well known to all working in the fields of forensic anthropology and bioarchaeology. The central issue for discussion in this context is criteria for recognizing the earliest evidence of bone response. A similar issue relates to how long prior to death fracture could occur without evidence of bone response.

The literature presents varied information on the timing of initial antemortem bone response to trauma but agrees that response is earlier in juveniles than in adults (Ubelaker and Montaperto, 2011). The published literature suggests that antemortem bone response generally can be detected between four and 15 days after the event. On the early end of this range, O'Connor and Cohen (1987) suggest that in juveniles, some periosteal new bone formation occurs as early as four days after fracture. They also report that such formation can be delayed up to 21 days. Most other estimates of the first appearance of bone response fall within this range. Maples (1986) indicated that the earliest evidence of antemortem response consists of subtle rounding of fractured bone margins with a polished appearance at the microscopic level. Mann and Murphy (1990) suggested two weeks for early onset of bone remodeling consisting of bone resorption on the periosteal bone surface near the fracture site.

7. Postmortem Occurrence of Patterns Usually Associated with Perimortem Conditions

The forensic anthropologist must exercise caution in diagnosing perimortem status from alterations frequently but not exclusively associated with perimortem trauma. Alterations suggestive of blunt force trauma, sharp force trauma and even projectile trauma can all potentially represent postmortem events.

Ubelaker and Adams (1995) present the following case in point. In 1993, a construction worker in Georgia discovered human skeletal remains during the process of clearing brush in preparation for house construction in the area. Although the remains represented one adult individual, the bones were found disarticulated, apparently disturbed by machinery working in the area. Analysis revealed the remains likely originated from a 25 to 35 year old female, approximately five feet seven inches tall and likely of European ancestry. Bones displayed evidence of past rodent gnawing and extensive taphonomic alterations suggestive of an extended postmortem interval.

In addition to many clear postmortem taphonomic alterations, the bones displayed triangular-shaped fractures of the long bone diaphyses of both humeri, the left tibia and left fibula. These triangular fractures are consistent in pattern with those described in the clinical literature as "butterfly fractures" (Harkess, 1975). Such fracture types are well known to be associated with traffic collisions involving pedestrians. When the automobile impacts the legs or other appendages of the pedestrian, the force operating on the affected long bone diaphyses creates compression stress and concave bending at the impact site and tension stresses on the opposite side. Fractures on the compression, impact surface tend to be complex and affect a relatively broad area. In contrast, fractures on the tension side tend to be simple. The net effect of these alterations can be a triangular shaped area of fractured bone with the base of the triangle on the compression side and the apex of the triangle on the tension side. This general pattern was present on the long bones from the Georgia case discussed here.

However, more detailed examination revealed distinct coloration differences between the general periosteal surfaces of the bones and the internal surfaces at the fracture sites. The external surfaces had been stained brown. In contrast, the internal bone cortex exposed at the fracture sites presented the natural cream color of bone. This coloration contrast indicated that the fractures had been sustained shortly before recovery. The alterations likely resulted from the trampling effect of the heavy power equipment operating in the area.

8. Moisture Content

Much of the diagnostic evidence for the perimortem condition (in contrast to postmortem) consists of the plastic response characteristic of fresh or green bone. As noted by Sauer (1998) the condition of dryness is the key variable and timing is environmentally dependent. Maples (1986) suggested that the fresh bone condition may persist for several weeks in an environment conducive to moisture retention. As early as 1977 Fitzgerald noted that "real materials are generally viscoelastic (or elastoviscous); that is they deform under mechanical loads partly as viscous liquids and partly as elastic solids." (p. 49). Thus, the moisture content of bone affects how it will respond to the mechanical loads involved in trauma.

In a relevant and innovative experiment, Weiberg and Wescott (2008) examined the variables of color variation, fracture morphology and microscopic characteristics of porcine long bones experimentally traumatized. This project noted how moisture content and the flexible collagen matrix of fresh bone play key roles in bone response to trauma. Long bones were tested for trauma response every 28 days throughout a 141 day postmortem period. They found a rapid change in bone moisture content and noted that the bone no longer retained fresh characteristics after 141 days. Throughout this period, fractures changed from initial smooth surfaces with obtuse or acute angles with curved or v-shaped outlines to jagged surfaces with more right angles and few curved or v-shaped outlines. Their study also revealed considerable variation in the preservation of fresh bone characteristics throughout the experimental period.

9. Summary

Although the concept of perimortem is a relatively recently used term in forensic science, it expresses the complexity of interpretation of the timing of trauma and other factors. Its meaning and intent varies among practitioners in different areas of medical science, arguing for caution and explanation when appropriate.

Within forensic anthropology, the term applies to interpretations regarding the condition of bone at the time of the trauma event. Thus the term relates to discussion of timing rather than the death event itself. The size of the perimortem window can vary considerably depending on the context and circumstances of a case. This variation is created by challenges in the differentiation of perimortem and postmortem events. Some practitioners use the term narrowly to refer to specific patterns and alterations suggesting fresh bone with high moisture content were involved. To others, it becomes a default category when alterations lack indications of antemortem and postmortem status.

Through case work and thoughtful experimentation, knowledge continues to grow regarding the complexity of perimortem alterations and their interpretation. Such advance is welcomed since trauma interpretation is such an important aspect of the many contributions of forensic science.

Acknowledgements

Kristin Montaperto, Keitlyn Alcantara and Dilliana Anaya-Ramirez of the Smithsonian Institution assisted with literature search and aspects of manuscript preparation.

References

Adams, B.A. 2007. Assessing trauma and time since death, in: Adams, BA. *Forensic Anthropology*. Infobase Publishing, New York, NY, pp. 50–64.

Berryman, H.E., Haun, S.J. 1996. Applying forensic techniques to interpret cranial fracture patterns in an archaeological specimen. *International Journal of Osteoarchaeology* 6, 2–9.

Dix, J., Calaluce, R. 1999. *Guide to Forensic Pathology*. CRC Press, Boca Raton, FL, pp. 138–139.

Dutelle, A. 2011. *An Introduction to Crime Scene Investigation*. Jones and Bartlett Publishers, Sudbury, MA, pp. 414.

Fisher, R.S. 1973. Time of death and changes after death, in: Spitz, W.U., Fisher, R. (Eds.), *Medicolegal Investigation of Death: Guidelines for the Application of Pathology to Crime Investigation*. Charles C. Thomas, Springfield, IL, pp. 23–27.

Fitzgerald, E.R. 1977. Postmortem transition in the dynamic mechanical properties of bone. *Medical Physics* 4, 49–53.

Fransioli, M.G., Szabo, E.T., Sunshine, I. 1980. Detection of methadone and propoxyphene in stored tissue. *Journal of Analytical Toxicology* 4(1), 46–48.

Galloway, A., Symes, S.A., Haglund,W.D., France, D.L. 1999. Trauma analysis, in: Galloway, A. (Ed.), *Broken Bones: Anthropological Analysis of Blunt Force Trauma*. Charles C. Thomas, Springfield, IL, pp. 5–31.

Haglund, W.D., Sorg, M.H. 2002. *Advances in Forensic Taphonomy: Method, Theory, and Archaeological Perspectives*. CRC Press, Boca Raton, FL.

Harkess, J.W. 1975. Principles of fractures and dislocations, in: Rockwood, C.A. Jr., Green, D.P. (Eds.), *Fractures*. J.B. Lippincott Co., Philadelphia, PA, pp. 1–11.

Herrmann, N.P., Bennett, J.L. 1998. The differentiation of traumatic and heat-related fractures in burned bone. *Journal of Forensic Science* 44(3), 461–469.

Krogman, W.M. 1962. *The Human Skeleton in Forensic Medicine*. Charles C. Thomas, Springfield, IL.

La Follette, L., Jacobson, L.B., Hill, J.D. 1980. Isolated aortocoronary bypass operations in patients over 70 years of age. *Western Journal of Medicine* 133(1), 15–18.

Loe, L. 2009. Perimortem trauma, in: Blau, S., Ubelaker, D.H. (Eds.), *Handbook of Forensic Anthropology and Archaeology*. Left Coast Press, Walnut Creek, CA, pp. 263–283.

Mann, R.W., Murphy, S.P. 1990. *Regional Atlas of Bone Disease*. Charles C. Thomas, Springfield, IL.

Maples, W.R. 1986. Trauma analysis by the forensic anthropologist, in: Reichs, K. (Ed.), *Forensic Osteology: Advances in the Identification of Human Remains*. Charles C. Thomas, Springfield, IL, pp. 218–228.

McManus, A.T., Mason, A.D., McManus, W.F, Pruitt, B.A. 1985. Twenty-five year review of *Pseudomonas aeruginosa* bacteremia in a burn center. *Journal of Clinical Microbiology and Infectious Diseases* 4(2), 219–233.

Merrill, G.K., Powell, R.J. 1980. Paleobiology of juvenile (nepionic?) conodonts from the Drum Limestone (Pennsylvanian, Missourian-Kansas City area) and its bearing on apparatus ontogeny. *Journal of Paleontology* 54(5), 1058–1074.

Moraitis, K., Eliopoulos, C., Spiliopoulou, C. 2009. Fracture characteristics of perimortem trauma in skeletal material. *Internet Journal of Biological Anthropology* 3(2).

Moraitis, K., Spiliopoulou, C. 2006. Identification and differential diagnosis of perimortem blunt force trauma in tubular long bones. *Forensic Science, Medicine and Pathology* 2(4), 221–229.

O'Connor, J.F., Cohen, J. 1987. Dating fractures, in: Kleinman, P.K. (Ed.), *Diagnostic Imaging of Child Abuse*. Williams & Wilkins, Baltimore, MD, pp. 103–113.

Passalacqua, N.V., Fenton, T.W. 2012. Developments in skeletal trauma: Blunt force trauma, in: Dirkmaat, D. (Ed.), *A*

Companion to Forensic Anthropology. Blackwell Publishing Ltd., West Sussex, UK, pp. 400–412.

Pechnicová, M., Porta, D., Cattaneo, C. 2011. Distinguishing between perimortem and postmortem fractures: are osteons of any help? *International Journal of Legal Medicine* 125, 591–595.

Pope, E.J., O'Brian, S.C. 2004. Identification of traumatic injury in burned cranial bone: An experimental approach. *Journal of Forensic Science* 49(3), 1–10.

Sauer, N.J. 1998. The timing of injuries and manner of death: Distinguishing among antemortem, perimortem and postmortem trauma, in: Reichs, K.J., (Ed.), *Forensic Osteology: Advances in the Identification of Human Remains*. Charles C. Thomas, Springfield, IL, pp. 321–332.

Scientific Working Group for Forensic Anthropology (SWGANTH). 2011. "Trauma Analysis". http://swganth.startlogic.com/Trauma%20Rev0.pdf (accessed 07/11/12).

Stewart, T.D. 1979. *Essentials of Forensic Anthropology*. Charles C. Thomas, Springfield, IL.

Symes, S.A., Dirkmaat, D.C., Woytash, J.J., *et al.* 2005a. Perimortem bone fracture distinguished from postmortem fire trauma: a case study with mixed signals. *Proceedings of the American Academy of Forensic Sciences* 11, 288–289.

Symes, S.A., Kroman, A.M., Rainwater, C.W., Piper, A.L. 2005b. *Bone Biomechanical Considerations in Perimortem vs. Postmortem Thermal Bone Fractures: Fracture Analyses on Victims of Suspicious Fire Scenes*. Poster presented at the Annual Meeting of the American Association of Physical Anthropology, Milwaukie, OR.

Symes, S.A., L'Abbé, E., Chapman, E.N., Wolff, I., Dirkmaat, D. 2012. Interpreting traumatic injury to bone in medicolegal investigations, in: Dirkmaat, D.C. (Ed.), *A Companion to Forensic Anthropology*. Blackwell Publishing Ltd, West Sussex, UK, pp. 340–389.

Symes, S.A., Rainwater, C.W., Chapman, E.N., Gipson, D.R., Piper, A.L. 2008. Patterned thermal destruction of human remains in a forensic setting, in: Schmidt, C.W., Symes, S.A. (Eds.), *The Analysis of Burned Human Remains*. Academic Press, Burlington, MA, pp. 15–54.

Turner, C.G. II 1983. Taphonomic reconstructions of human violence and cannibalism based on mass burials in the American Southwest, in: LeMoine, G.M, MacEachern, A.S. (Eds.), *Carnivores, Human Scavengers & Predators: A Question of Bone Technology*. Proceedings of the Fifteenth Annual Conference, the Archaeological Association of the University of Calgary. University of Calgary Archaeological Association, Calgary, AB, pp. 219–240.

Turner , C.G. II, Turner, J.A. 1990. Perimortem damage to human skeletal remains from Wupatki National Monument, Northern Arizona. *Kiva* 55(3), 187–212.

Ubelaker, D.H. 1991. Perimortem and postmortem modification of human bone. Lessons from forensic anthropology. *Anthropologie* XXIX/3, 171–174.

Ubelaker, D.H., Adams, B.J. 1995. Differentiation of perimortem and postmortem trauma using taphonomic indicators. *Journal of Forensic Science* 40(3), 509–512.

Ubelaker, D.H., Montaperto K.M. 2011. Biomechanical and remodeling factors in the interpretation of fractures in juveniles, in: Ross, A.H., Abel, S.M. (Eds.), *The Juvenile Skeleton in Forensic Abuse Investigations*. Humana Press, New York, NY, pp. 33–48.

Van Hoof, J. 1979. Influence of ante- and peri-mortem factors on biochemical and physical characteristics of turkey breast muscle. *Veterinary Quarterly* 1(1), 29–36.

Weiberg, D.A.M., Wescott, D.J. 2008. Estimating the timing of long bone fractures: Correlation between the postmortem interval, bone moisture content, and blunt force trauma fracture characteristics. *Journal of Forensic Science* 53(5), 1028–1034.

11. You Are What You Ate: Using Bioarchaeology to Promote Healthy Eating

Jo Buckberry, Alan Ogden, Vicky Shearman and Iona McCleery

The You Are What You Ate *project is a collaboration between historians, archaeologists, museum officers, medieval re-enactors and food scientists. We aim to encourage public debate and personal reflection on modern eating habits through exploration of the dietary choices of the medieval and early modern period. This paper will discuss our osteology workshops, aimed at adults or at school children.*

We use archaeological examples of diet-related conditions, including dental disease, scurvy, rickets and gout, plus those associated with obesity such as osteoarthritis and DISH, to help the public visualise how dietary choices can affect the body. This information is delivered via an introductory talk and carefully monitored bone handling sessions, and, for the children, includes the analysis of a plastic skeleton modified to display pathological conditions.

Evaluation data shows that the majority of adults and all children feel they have learnt something new during the sessions, and that this has led them to think about healthy eating. The inclusion of examples of dental pathology has promoted dental hygiene to school children, although it was not one of our primary aims. It is difficult to assess if these short-term experiences translate to long-term knowledge gain or to changes in behaviour.

Keywords: Public engagement; Osteology; Workshops; Diet; Medieval; Healthy eating

1. About You Are What You Ate

The *You Are What You Ate* project (hereafter YAWYA; http://www.leeds.ac.uk/yawya/) is a collaboration between historians, archaeologists, osteologists, museum officers, food scientists and medieval re-enactors. We are supported by the Wellcome Trust through a Society Award (2009 themed call on eating). Our project aims to promote debate and personal reflection on healthy eating by exploring dietary choices of the past, specifically the medieval (here defined as the 13th to 15th centuries) and early modern (16th and 17th centuries) periods. Our project is based in West Yorkshire and, due to the collaboration with the museums and museum education service of Wakefield Council, we targeted the Wakefield district specifically. We focussed on the medieval period because of the expertise of team members combined with local interest in the Battle of Wakefield (AD 1460), Sandal Castle and the curation of a large number of medieval skeletons at the University of Bradford, including many local to the district. The inclusion of the early modern period allows us to discuss changes in diet that a more global economy introduced and allowed us to use Clarke Hall, Wakefield (built AD 1680) as a venue for cooking workshops. We have deliberately designed a range of activities and events to reach as wide a demographic as possible. While some events will reach a wide audience, others are designed and run to cater to specific groups (adults, children, those already interested in archaeology, history, or food and eating), some of which are considered 'hard to reach' including 13–19 year olds, young people not in education, employment or training (NEETS), young people at risk of exclusion, Black and Minority Ethnic (BME) young people and young people with learning difficulties. The three-year project (running 2010–2013) includes a wide range of public engagement activities: stalls at food festivals and local markets, museum exhibitions, school assemblies and activities, food workshops for children and adults, youth workshops, osteology workshops for both adults and children, evening talks and 'historic food days' (public research conferences).

A key theme for YAWYA was how diet affects the body. Working with food scientists, we wanted to stress that no food was inherently bad, but that getting the right balance can improve our health, thus we focussed on metabolic diseases and the effects of both obesity and undernourishment on the skeleton in particular. We believe that showing how bodies are affected in a visually stimulating environment has the potential to make a bigger impact on people than merely stating something is 'good for you' or 'bad for you'. Anecdotally, the first two authors (JB and ARO) were aware that students (and staff!) tend to modify their behaviour slightly following certain palaeopathology labs; the desire to clean teeth following anything on dental disease, or an increased awareness of poor posture after examining spinal joint disease, for example. Specific issues were addressed in our first two exhibitions: 'Sugar and Spice and All Things Nice' (Wakefield Museum, March to September 2011) included a section on the relationship between sugar consumption and the prevalence of caries in the past (Moore and Corbett, 1978), whereas 'The Dark Side of Eating' (Pontefract Museum, March to September 2012) discussed a wide range of diet-related pathologies including obesity (and the increased risk of osteoarthritis, especially of the knee), gout, scurvy and rickets (Roberts and Manchester, 1995). We drew on these themes for the osteology workshops, which were timed to coincide with the exhibitions. This paper will discuss the osteology workshops in more detail, highlighting our aims, session structure/content and the results of our evaluation.

2. About the Workshops

In our initial plan, we aimed to deliver two osteology workshops to adults, and two to Key Stage 2[1] school children per year; after the success of the first year we expanded the provision of children's osteology workshops to four per year. At the time of writing we had delivered four adult workshops and six children's workshops, engaging with 94 adults and 141 children plus 22 teachers/accompanying adults. All of the workshops were planned to include two activities – a short talk followed by a hands-on workshop – each modified to meet the needs of the audience and/or the exhibition theme. It is well known that people are naturally predisposed to different learning styles (Hein, 1998). By combining auditory, visual and kinaesthetic components in each workshop we hoped to make the sessions memorable to as many people as possible. For the adults, the hands-on workshop offered the opportunity to examine real archaeological material; this was preceded by a brief introduction about the ethics of excavating and researching human remains and appropriate behaviour. For the children, a series of carefully selected archaeological specimens were shown to small groups of children by demonstrators, who facilitated discussion about the pathology present and the possible dietary causes. This was followed by an activity known as 'skeleton in a box' where, working in small groups, the children laid out and analysed a modified plastic skeleton (see section 2.2.1).

2.1 The Aims of the Adult Workshops

We anticipated that the adult osteology workshops might attract repeat participants; therefore we decided to address slightly different themes each year. In the first year we focussed on dental disease as this linked well with the exhibition 'Sugar and Spice and All Things Nice'. The first two workshops were identical, and lasted 2.5 hours. One was on a Wednesday and one on a Saturday; in later years we decided to run all of these events on Saturdays as fewer people were able to attend the Wednesday session and thus attendance was lower. The aims for the first two adult workshops were:

1. To engage the lay public in human osteology
2. To show the effects of diet on teeth and jaws
3. To compare patterns of medieval oral health with modern oral health, and to see what lessons can be learned
4. To raise awareness of the ethical treatment of working with human remains

In the second year, we expanded the osteology workshops to also include skeletal pathologies, taking the lead from our exhibition theme 'The Dark Side of Eating'. Thus our second and third aims were modified to include skeletal diseases, increasing the pathologies to encompass scurvy, rickets, cribra orbitalia, osteoarthritis, diffuse idiopathic skeletal hyperostosis (DISH) and gout (Roberts and Manchester, 1995; Brickley and Ives, 2008). The first workshop was run as part of the British Festival of Science (in Bradford) in September 2011 and was limited to two hours; the second was timed to occur shortly after 'The Dark Side of Eating' opened in Pontefract, and reverted to the 2.5 hour format.

For the first half of each workshop ARO delivered a lecture which illustrated the different pathologies and highlighted their aetiologies and especially links to diet. Where possible, observations were made about the presence of these diseases in modern populations and, for some (e.g. dental caries), comments were made about differences in treatment. After a short break participants were taken to the osteology lab, where they were each given a lab coat. We briefed them on the ethical treatment of human remains, encouraging them to learn from the remains laid out in the lab but to handle them with care, if at all. A team of five demonstrators, including JB and ARO, circulated around the lab discussing the specimens and the observable pathologies. In each case we tried to explain the aetiologies and the links to diet, but stressed that these were complex. Invariably these informal discussions developed to include personal observations regarding participants' own health, or those of their friends and family, and topical subjects including the fact that rickets is being diagnosed more often in recent years in British populations (Pearce and Cheetham, 2010), discussed as a topical news item on the YAWYA web site (McCleery and Buckberry, 2010). At the end of the session, all participants and team members were asked to complete an evaluation form.

2.2 Children's Workshops

The osteology workshops for children were designed to suit a range of different learning styles; visual, kinaesthetic, auditory and problem-solving, allowing all children to learn in the way that suits them (Smith, 1998). The workshops included repetitive elements, delivered in different ways (talk, show-and tell, discussion, hands-on activity). The activity encouraged teamwork and discussion, which has been shown to be particularly beneficial for girls (Askew and Ross, 1988; Fennema, 1996), whereas boys have been shown to need physical interaction in order to learn and therefore they often respond well to hands-on activities (Russell, 1994). At the end of the session each group reported back to the class, promoting and developing public speaking and presentation skills. Regular discussions with the large team of demonstrators, and teachers and classroom assistants accompanying each group allowed pupils to express their own views throughout the hands-on activity.

The children's workshops were advertised through Wakefield Council's iPoint school service and via the museum's school contact list, and were made available to Key Stage 2 children, but were specifically targeted to Year 6 (the final year before secondary school, ages 10 to 11 years). Each session was for one class of approximately 30 children, with accompanying teachers and adults. In line with the adult workshops, during the first year we focussed on dental disease. We expanded the workshop to include other conditions in the second year of delivery. The aims of the children's workshops were:

1. To engage school children in human osteology
2. To show the effects of diet on teeth and jaws
3. To show that similar dental diseases are seen in both archaeological and modern populations, the main difference is in the level of medical care available.
4. To promote the *You Are What You Ate* project to teachers

In the second year, the second and third aims were modified to reflect that the focus was now on the entire skeleton, not just the dentition.

The school groups were bussed from Wakefield to Bradford. On arrival, the children and teachers were taken to our osteology teaching lab where they were given lab coats to wear; as well as protecting clothing, this reinforced the health and safety messages (don't eat, wash your hands), but also added to the sense of occasion. We started by delivering a short talk on how and why we analyse skeletons, focussing on the diet-related pathologies and their causes. It was a challenge to pitch these to the right audience, but by using humorous images and asking lots of questions, and keeping this as short as possible, most of the children appeared to be engaged throughout the talk. All of the demonstrating team were present during the talk, as it allowed us to ensure we were consistent in the messages that we were conveying.

After the talk, each demonstrator selected one archaeological specimen with a different pathology. We circulated around the room, talking to the children (and adults!) in small groups about what we could see, what the pathology looked like, what disease this was, and which type of food might have caused the condition, and what type of person (rich, poor) might be affected. We asked the children questions (linking back to the talk to reinforce the content) and encouraged as much discussion as possible. If children wanted to touch the remains we allowed this, but only within the constraints of this highly controlled environment. Once all of the children had seen each example, we started the main activity 'skeleton in a box', modified from school sessions which have been run by John McIlwaine in Bradford for the last ten years. In the YAWYA activity, we used modified plastic skeletons, altered to show evidence of the diseases discussed in the first half of the session.

2.2.1 Modifying Plastic Skeletons

Ten commercially available, low price full size plastic replica skeletons (described as a 'full disarticulated budget skeleton with skull') were purchased. In the two weeks prior to the first course they were modified to show clear evidence of a range of diet-related dental pathologies, including dental caries, dental granulomas and abscesses, dental calculus and ante-mortem tooth loss (Ogden, 2008). Prior to the workshops in the second year, these skeletons were further modified to show clear evidence of diet-related skeletal pathologies including rickets, cribra orbitalia, scurvy, DISH and gout (Roberts and Manchester, 1995). As in the real world, some skeletons had multiple pathologies. Three principal methods were used, which are discussed below.

Building-up

The surfaces to be built-up were cleaned, dried and slightly roughened with file strokes to ensure a good bond. An ivory-coloured fast setting epoxy putty (Milliput©) was then mixed according to the manufacturers guidelines. Like most chemicals in domestic use this product may cause irritation to sensitive skins, and when kneading the two components together it is advisable to use disposable gloves. The putty is sticky and adhesive when first mixed and can then be placed where needed. After 30 minutes it becomes rubbery and less tacky. After about one hour it can be cut into shape with a knife. During this period it can be shaped by modelling tools, and its surface smoothed

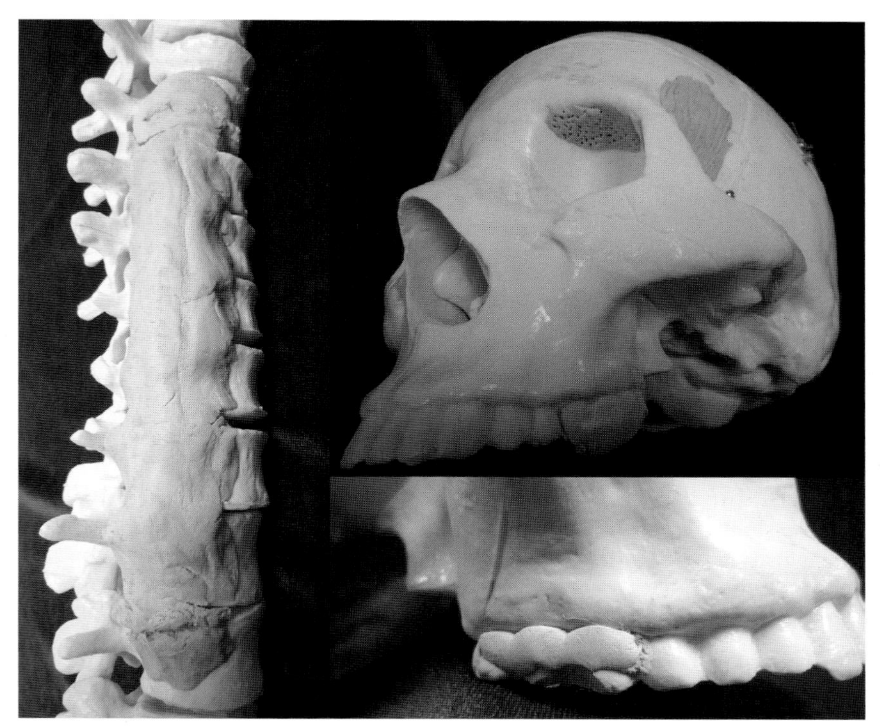

Figure 11.1. Examples of 'build-up' pathology on the modified skeletons. Left: DISH; top right: scurvy; bottom right: calculus.

with moistened tools or wet gloved fingers. The material becomes hard after four hours at 20–25°C. It can then be filed, sawn or even painted to improve realism.

By these means we were able to simulate calculus deposition on the teeth, and the bone deposition in the orbits and beneath the muscles of mastication, seen in scurvy, and the flowing spinal osteophytes characteristic of DISH (Figure 11.1).

Drilling

In a well-ventilated room (or even under extraction or a fume cupboard, if available) and wearing a face mask, a 12V hand-held drill (Proxxon Micromot ©) and 1–2mm diameter dental rose-head and tapered cross-cut burrs was used to drill shallow pores in the roofs of the orbits in a pattern suggestive of cribra orbitalia. Para-articular scooped out lesions, characteristic of gout, were drilled into the distal end of first metatarsals. Carious cavities were drilled into teeth, producing a range of sizes of occlusal and interproximal lesions. Where carious lesions were extensive, pulpal exposures and alveolar granuloma/cysts and abscesses in the alveolus were also fabricated. By this means teeth could also be removed and the alveolus 'resorbed' to simulate periodontal disease, anterior tooth loss due to scurvy and the long-term effect of posterior tooth loss (Figure 11.2).

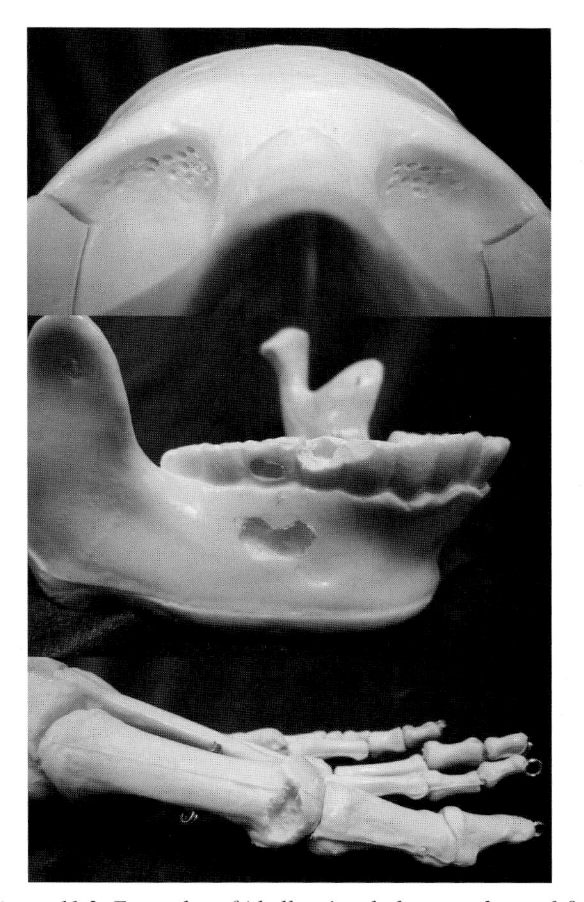

Figure 11.2. Examples of 'drilling' pathology on the modified skeletons: Top: cribra orbitalia; middle: caries and apical granuloma; bottom: gout.

Bending

In a well-ventilated room, extremely cautious softening of the mid-shafts of long bones with a 1500W hot-air

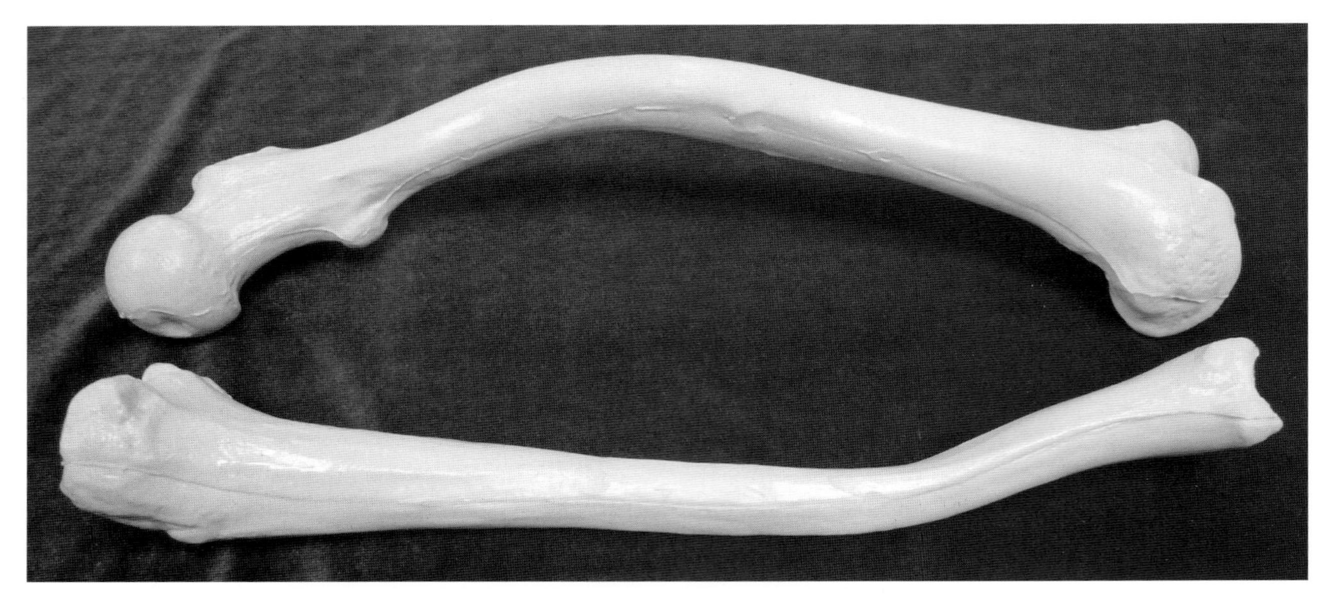

Figure 11.3. Rickets created by bending replica femur and tibia.

gun (for DIY paint-stripping), enabled the gentle bending of the shafts. The ends of the bones were then held firm in the new position whilst the 'bone' was cooled in cold water. By this means we were able to simulate the results of rickets in childhood on the long-bones of upper and lower limbs (Figure 11.3).

2.2.2 The Skeleton in a Box Activity

The children were organised into groups of 3 or 4 (with the help of the class teacher; in some cases the teachers deliberately moved the children about into specific groups). Each group was presented with a box containing a skeleton for analysis and each child had a specially designed recording form and YAWYA pencil. We asked the children to start by laying out their skeleton in anatomical order, using four hanging skeletons as a reference with the help of the demonstrator team and the accompanying teachers/ adults, and colouring in which bones were present. Next, they assessed the sex of the skeleton, using a small number of standard traits of the pelvis and skull (Buikstra and Ubelaker, 1994) and the age of the skeleton by observing the pattern of dental wear (Brothwell, 1972) and the morphology of the pubic symphysis (Brooks and Suchey, 1990). The images used for this section of the recording form were those used by professional osteologists (to keep this as authentic as possible) and were simplified to reduce the amount of specialist information. After the first year, the images were simplified further to ensure the children were not confused by having too much information available, based on staff observations and experiences in the first two sessions. Although we encouraged the children to use these methods as accurately as possible, if the group decided the skeleton was female (the replica skeletons were all males), we agreed with their assessments. Once age and sex had been established, the children were asked provide a name for the skeleton. This was added to the activity as we felt it would encourage the children to think about the skeleton as a person. The final part of the activity was to describe and diagnose the pathology present, with space to draw the pathology if the child wished to do so. Questions were then asked about what might have caused the pathologies; focussing on diet, but also incorporating lack of dental hygiene, as most of the jaws observed (both real and plastic) had evidence of calculus in particular. At the end of the session each group presented their findings to the rest of the class, and then packed their skeletons away.

After the session was completed, all participants (children, accompanying adults and – a little later on – the project team) completed evaluation forms. We then made sure everyone returned their lab coats, washed their hands, and were given a YAWYA freebie as they left: project stickers and either a project water bottle or set of coloured pencils.

3. Measuring Success – Evaluation of Events

Evaluation forms were designed with the help of cultural officers at Wakefield Museums and the project administrator, Fiona Blair. They were designed to contain both open and closed questions and therefore produce both quantitative and qualitative data. Questions were written to enable the analysis of Generic Learning Outcomes (GLOs) and Generic Social Outcomes (GSOs), devised by the Museums, Libraries and Archives Council (MLA), and used as a standard museum evaluation tool. This methodology recognises that learning is not just about knowledge transfer, but can also include inspiration, change or progression in attitude or behaviour, and new skills

Table 11.1. Responses to yes/no questions by adult participants (n=94).

Question	Yes	No	Yes and No	Don't know	Blank
Would you recommend this event to others?	92			1	1
Did you think that the event was well planned?	92		1		1
Have you learnt anything new from this event?	92	2			
Would you describe your participation in the workshop as a positive experience?	92	1			1
Do you think that it is okay for people to study and handle human bones from archaeological sites?	86		1	4	3

(Foster, 2008; Museums, Libraries and Archives Council, 2008). We have adapted this approach to investigate if we met our aims, and will undertake formal GLO and GSO evaluation of the data at the end of the project.

Four different audiences evaluated the workshops: adult participants; children; teachers (and other accompanying adults); and project team members. Questions explored the running of the event, if participants had been to another YAWYA event, enjoyment, the learning experience and about the ethics of studying human remains (adults only). For the school children, the final section prompted them to discuss 'what I will remember most – write or draw it', next to a large speech bubble.

All responses were collated onto excel spreadsheets, allowing us to collate qualitative comments and analyse quantitative data. All comments were typed up as written on the evaluation forms and any spelling mistakes were transcribed directly. Occasionally it was difficult to read handwritten comments. Comments selected for inclusion here were thought to reflect the main issues raised by the participants; occasionally comments which were contrary to general opinion were selected to demonstrate that differences of opinion were evident on the evaluation forms. It is hoped that these represent a sample of typical comments, while not being limited to just those who had either a positive or negative experience. Comments were also used to help develop the sessions further: team members were asked to reflect on how well sessions were run, how we had achieved our aims, and how we could improve similar sessions in the future. Comments by adult participants and teachers allowed us to develop ideas and to deal with small snags we had not identified ourselves. Thus each session evolved from the last, and evaluation forms were modified to avoid ambiguity or changed focus in subject matter. Chi-squared tests were undertaken to compare agree/strongly agree and disagree/strongly disagree, or yes/no responses using SPSS version 20.0. For the schools workshops, comparisons were made between boys and girls for yes/no responses to see of the sex of the participant had any effect on their experience. It is commonly reported that boys and girls have different levels of achievement at different developmental stages (Younger *et al.*, 2005), and we wanted to ascertain whether they had engaged with the workshop in a similar way. Chi-squared tests comparing yes/no responses for boys and girls were undertaken on 2×2 tables, and where appropriate Fisher's exact tests were used. The evaluation forms were all copied and are archived at the University of Leeds, with a second copy at the University of Bradford.

4. Results – Informal Observations and Formal Evaluation

Overall, the participants indicated both formally (via evaluation forms) and informally (by chatting to the team, or to each other) that they had had a positive, enjoyable and educational experience. After each event, the osteology team noted things that could be done differently, and incorporated these into our event plans for the next workshop. It was gratifying to note that after each event, the team evaluation became more positive, especially in terms of the event organisation. Actively reflecting on how things had gone helped improve the events.

4.1 Workshops for Adults

All of the quantitative data is strongly skewed towards the positive responses. Tables 11.1 and 11.2 collate the responses to a series of questions by adult participants based on yes/no type questions and those on an ordinal scale respectively. Almost all found the workshops had been well planned and that they would recommend the event to others. Only 2 participants felt they had not learnt anything new, and all who responded found the workshop to be 'interesting' or 'very interesting'.

The majority of participants (n=86, 91.5%) felt it was alright to study and handle archaeological human bone; the remaining individuals either responded 'don't know' or left the question blank. One participant circled both 'yes' and 'no', and qualified this with the statement:

> "Handling remains in a situation similar to today or as part of research is okay but not if you are not in an academic/scientific situation".

This positive response is unsurprising, as individuals opposed to the study of human bone are unlikely to have signed up to take part in this event.

Table 11.2. Responses to questions on an ordinal scale by adult participants (n=94).

How interesting did you think the workshop was?	Very interesting	Interesting	Okay	Not interesting	Boring	Don't know
	84	10				
In your opinion, was the overall standard of the workshop:	Excellent	Good	Satisfactory	Poor	Blank	
	76	17			1	

Each of the quantitative questions was coupled with an open, qualitative question (such as 'why do you say this?'). The responses to these questions were assessed by JB and VS. 61 participants answered 'Would you recommend this event to others? Why do you say this?', although sadly we could not read one response. All of the comments were positive, many highlighting that they had found the event interesting, that the lecture was both interesting and fun, and that all of the team were knowledgeable; indeed access to academics who enthused about their subject seemed to be a real draw. Several participants observed how important it was to be aware of health-related issues, and that it made them aware of modern-day issues. Others noted that they enjoyed the hands-on aspects of the workshop, and it was better seeing the bones than just photographs. Comments included:

> "It's generally really interesting and fun. It's applicable to everybody",
> "Interesting, informative, close up view of specimens",
> "Very accessible & introduced the subject on a friendly basis",
> "Very interesting presentation – reminding us of how DIET really impacts on health" (participant's emphasis),
> "Because it makes you understand why we do the things we do today to prevent these diseases",
> "Sheer quantity of freely given expert analysis. Very fluent, eminently understandable".

Many of these themes were picked up again in answers to the questions 'What did you particularly like about this workshop?' and 'Would you describe your participation in the workshop as a positive experience? Why do you say this'.

When we asked what people had learnt the answers were wide and varied, from great amounts of specific details to simple statements such as "A lot!". Each statement was read by JB and a tally was kept of how many times comments reflected certain key themes: health/disease/lesions (n=31), care/treatment of conditions (n=7), comparisons of the past and the present (n=3), references to diet (n=10), anatomical knowledge (n=17), scientific approaches/techniques of analysis (n=5) and personal reflection (n=3). One comment could easily encompass several of these broad themes. Anecdotally, it was obvious from talking to the participants that most of them had learnt something about most of these areas, but of course they would only write a fraction of that down on the forms.

The majority of people made a comment about seeing what different pathologies looked like, understanding disease processes more or made a comment about a specific pathology, an expected outcome given the focus of our workshop, and indicating we had successfully raised their awareness of human osteology (Figure 11.4). Some comments included personal reflection, suggesting that the workshop had encouraged personal reflection on individual but also family health:

> "The effects of vitamin D on rickets and how it has affected my family history",
> "Names of tooth and gum problems. Thankful that I have dentures!".

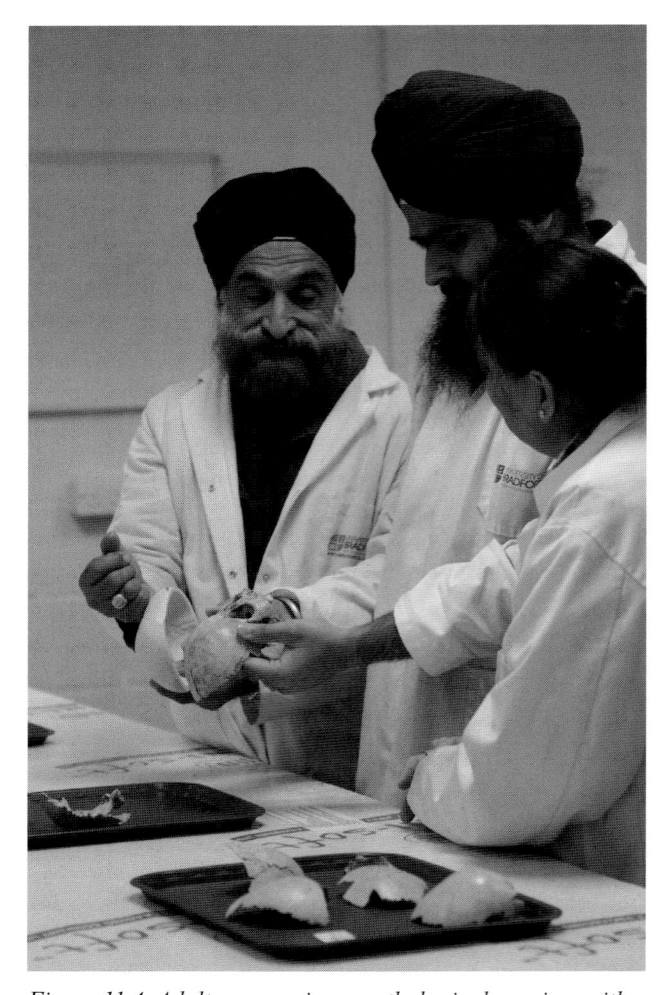

Figure 11.4. Adults comparing a pathological cranium with a non-pathological cranium at a YAWYA osteology workshop.

Interestingly, most of the demonstrating team had had lengthy discussions with participants about family and friends with certain diseases; in particular the fact that gout is in part hereditary seemed to resonate with those who knew people with gout running through their families. General observations on the level of health were common, but those that directly compared past and present health were rare:

> "How terrible dental health was in the past and also how living now brings it's (sic) own problems".

It was interesting to see that many people learnt some basic anatomical knowledge, and clearly appreciated being able to see and handle real bones:

> "DISH and what a normal spine looks like! Never seen a whole skeleton laid out before",
> "Effects of vitamin deficiency etc. on bones. What human bones really look and feel like",
> "I know a bit but seeing the actual material really made a difference".

Finally, several comments noted that participants had learnt about *how* we analyse skeletons and identify the diseases; becoming aware of the scientific process itself was an important learning outcome:

> "How evidence can be found",
> "Effect of evolution of diet on disintegration of teeth and methods of archaeological investigation".

It was clear that the theme of the workshop influenced the types of comments made; those from the first two workshops, linked to 'Sugar and Spice and All Things Nice', tended to specifically mention teeth, whereas those in the following year, when we took a much broader approach, tended to be far less specific.

4.1.1 Evaluation of Aims – Adult Workshops

Before the workshops we set specific aims; each will be discussed here. By attending the workshops, each participant had become aware of human osteology, which was our first aim. Evaluation indicated that they had really enjoyed the hands-on aspect of the workshop. The interaction with multiple demonstrators really enhanced that experience:

> "Hands-on! Real bones. Really old bones. Amazing" (participant's emphasis),
> "Examining skeletons and meeting friendly people",
> "The opportunity to handle the teeth, with knowledgeable demonstrators who could provide extra information".

Our second aim was to show the effects of diet on teeth and jaws/on the skeleton and dentition. 97.9% of participants indicated that they learnt something new at the event, and the subsequent comments indicated this was often connected to the specific appearance of skeletal/dental lesions:

> "The mechanism of the diseases. How disease leaves a trace on your bones...",
> "How different diets can affect different aspects of the skeleton".

Our third aim was to compare patterns of medieval (oral) health with modern (oral) health, and to see what lessons can be learned. Some participants commented on the contrast between modern and medieval health:

> "That people's health was worse than I thought it was and how easy we have it today".

We did not specifically set out to change people's behaviour (and this would be very difficult to evaluate), but some comments suggested that seeing this evidence encouraged participants to reflect on their own habits:

> "To improve my diet and brush my teeth more often",
> "I'm reflecting on my diet as a result!".

Others recognised the need to pass this information on to a wider audience:

> "...we need to focus on telling people how much a poor/unbalanced diet impacts on their health".

Our final aim was to raise awareness of the ethical treatment of working with human remains. Team members commented on their evaluation forms that the briefing section at the start of each hands-on session really emphasised this point. Participants often had informal discussions about this with team members, discussing their own opinions, and recognising the importance of skeletal data (although this is, of course, a self-selecting audience). Answers to the question 'Do you think that it is okay for people to study and handle human bones from archaeological sites? Why do you say this?' were generally in support of analysis (91.5% of responses answered 'yes'); this reflects the general trend for public support on the analysis of human remains revealed in a recent ICM opinion poll commissioned by English Heritage (Payne, 2010). Comments on YAWYA evaluation forms revealed that participants felt this was OK, provided they were treated appropriately:

> "It is the only way we can learn. Human remains hold no special significance to me but we must respect the sensibilities of others",
> "As long as it is done carefully and respectfully it is fine",
> "We'll only learn for the future by looking at the past as long as treated with respect it's fine",
> "With respect".

One participant clearly found it difficult to reconcile the educational and research value against the disturbance of graves:

> "I agree that you should examine them to learn but I also find it disrespectful to dig out peoples (sic) resting places. So I'm not sure!".

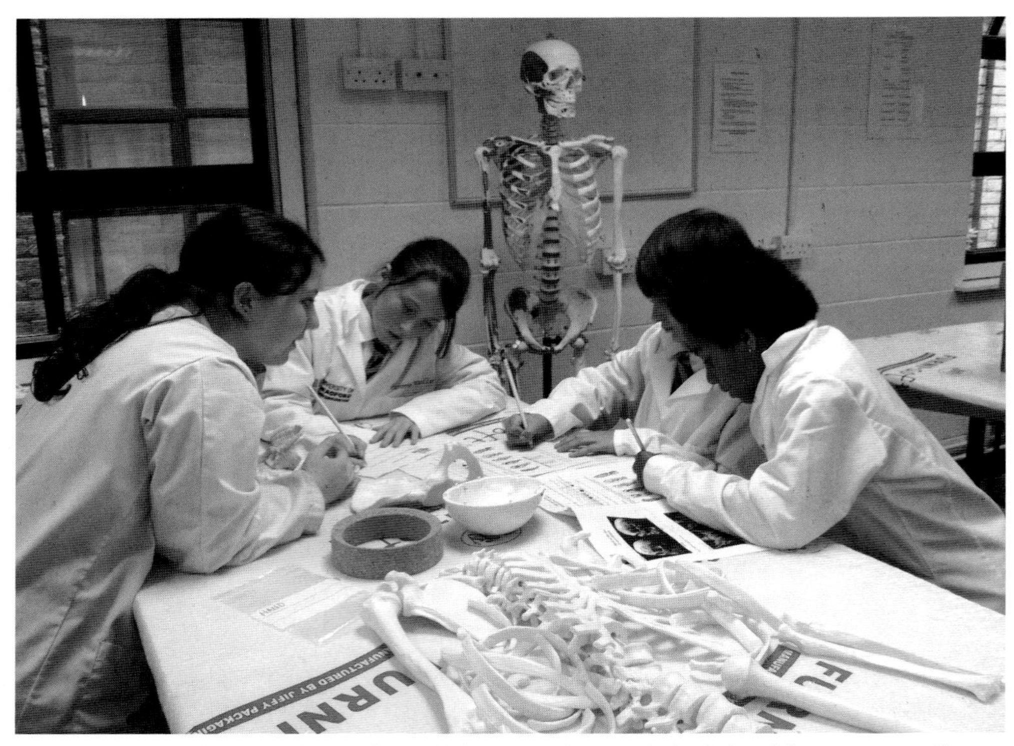

Figure 11.5. Children analysing their 'skeleton in a box' with the help of demonstrator Alice.

4.2 Workshops for Schools

This section combines quantitative data, qualitative data and starts with general observations by team members. It was clear from the outset that most of the children thoroughly enjoyed the workshops; most were clearly engaged with the event from the moment they donned a lab coat. During the talk they were generally attentive, and keen to answer questions, although this did vary between schools. Most groups who visited in the second year of the project had prior knowledge of scurvy, which they had encountered during the Key Stage 2 topic 'Tudor explorers', so when we asked them 'what kind of people get scurvy?' the answer was usually 'sailors' or 'explorers'. One unexpected, but technically correct, answer was 'pirates'! The team became aware that we were reinforcing the key messages delivered during the talk while we were discussing the archaeological bones in small groups, and prompted the children to remember key facts during this part of the session.

'Skeleton in a box' was clearly a very successful activity, and the children seemed to enjoy it immensely. When asked 'what did you enjoy most, write or draw it', many children described putting the skeleton back together, or drew a skeleton laid out on the table. Some groups became very engrossed in the puzzle-like challenge of getting all of the bones in anatomical order, going to great lengths to sort out the ribs and to articulate the loose hand and foot bones (each skeleton had one articulated and one disarticulated hand and foot), grasping the need to arrange this as a mirror image very quickly. During the first few workshops one team member observed that this seemed to be more common in groups of boys than groups of girls or mixed-sex groups, however in later sessions this task was completed by both boys and girls. Some groups became so engrossed in re-articulating the hands and feet that we had to encourage them to move on to the other aspects of the activity.

Many children were very good at assessing the age and sex of their skeleton, debating within the group what age/sex they thought the skeleton was, and supporting their observations with information and pictures provided on the worksheets (Figure 11.5). They all seemed to really enjoy thinking of a name for their skeleton, but some were a little bit perplexed that they had to make this up, and not work it out from some data provided by us, perhaps revealing how immersed the children were in the scientific scenario. This was the one aspect of the activity that seems to spark the greatest debate within the groups, and caused the most hilarity for team members. As a compromise, many of the skeletons had multiple middle names.

The children spotted most of the pathologies very quickly. Interestingly, the more subtle changes – those of the dentition, orbits and feet – were not equally easy to find. The scooped out lesions of gout were most commonly overlooked (until prompted to look at the feet), whereas equally small lesions in the cranium were generally observed early on, perhaps due to the focus of attention of the skull and dentition while assessing age and sex. Rickets was over-diagnosed by all groups; due to the inexpensive nature of the skeletons, quite a few had curved fibulae.

Table 11.3. Responses to yes/no questions by participants at schools workshops (n=141).
**55 children from two schools were not asked this question, therefore n=86.*

	Yes	No	Don't know	Blank
I had fun today	140			1
I learnt some new things	141			
I learnt some things that surprised me about diseases and eating	130	8		3
This was an exciting activity	138	1		2
I want to find out more	129	11		1
Do you think medieval people had healthier diets than us?*	22	63		1
I am interested in learning about the past	129	8		4
Today has given me some ideas about how to eat more healthily	128	11		2
Today has given me some ideas about looking after my teeth	138	2		1

Table 11.4. Comparison of responses to yes/no questions by boys and girls at schools workshops (n=141).
**55 children from two schools were not asked this question, therefore n=86*
Chi^2 tests were performed using SPSS 20.0 to compare responses (yes or no) of boys and girls. All tests bar ^ were Fisher's exact tests, due to low cell counts. P-values below 0.05 were considered significant.

	Boys		Girls		Chi^2 p-value
	Yes	No	Yes	No	
I had fun today	52		58		-
I learnt some new things	52		59		-
I learnt some things that surprised me about diseases and eating	49	1	52	6	0.120
This was an exciting activity	52		58		-
I want to find out more	49	2	55	4	0.684
Do you think medieval people had healthier diets than us?*	5	27	9	20	0.153^
I am interested in learning about the past	48	3	55	2	0.665
Today has given me some ideas about how to eat more healthily	49	2	51	7	0.170
Today has given me some ideas about looking after my teeth	51	1	57	1	1.000

This suggests the children had understood the diagnostic features of rickets and were noticing the mild curvature of the fibulae as well as the severe changes we had produced in the tibiae and femora. Generally speaking, the children were very observant, and remembered what the pathologies were. Some found it more difficult to associate the skeletal lesions with diet compared with dental lesions. Perhaps the general awareness that eating sweets caused holes in your teeth and the importance of dental hygiene contributed to this pattern.

The pupil presentations given at the end of the lab were often of an incredibly high standard, with many groups justifying their age and sex assessments and their diagnoses with observations. Where there was disagreement over the sex of the skeleton due to disparity between the skull and the pelvis, some groups chose the sex as assessed from the pelvis, because they had learnt that this was more accurate. This is important, as it reveals an understanding of the analytical process.

Evaluation data showed that every child indicated that they had learnt something new, and 130 children (92.2%) indicated that they had learnt something about diseases and eating (Table 11.3). Comments indicated that they had learnt how to lay out and analyse a skeleton:

> "I can tell if the skull is a boy/girl by the eyebrows",
> "how to bild a scealton".

Others clearly identified specific diseases in their answers, often using specialised language:

> "Your legs bend if you don't get enough sunlight, the disease is called rickets",
> "About granulomas",
> "I didn't know you could get serious desises just by not eating the right amount of iron or vitamin C/D",
> "scurvy makes you bleed inside".

In some cases, the general message was clearly understood, even if the detail was a little inaccurate:

> "That if you don't eat your greens then you will get holes in your brain"!

All bar one child (who did not respond to the question) had enjoyed the workshop and 138 (97.9%) thought the activity was exciting. Many of the teachers indicated they thought the activity was positive and one teacher commented (verbally) that it was good to see an activity that was equally enjoyed by boys and girls. Indeed, chi-squared tests revealed no significant differences at the 95% confidence level in the responses of girls and boys for any of the quantitative questions (Table 11.4). When asked what surprised them about the workshop, some children noted that:

> "It wasn't like a normal school lesson",
> "I didn't think we would've done things that were as fun as they were",

"In normal lessons we don't join bones together!",
"Because we never get to do owt like that in class" and
"That we get to wear lab coats".

This indicates that the session as a whole was a very different experience (and perhaps more fun) compared to school. Most of the children wanted to find out more and were interested in learning about the past (129, 91.5% for both questions).

Of the 86 children asked if medieval diets were healthier than modern diets, 22 (25.6%) thought they were and 63 (73.3%) thought modern diets were healthier (Table 11.3). The reasons given for this were varied, but the lack of sugar in the past was a common theme. Some children noted that medieval peasants had less access to food:

"Because they were poor and they didn't have as much sugar",
"because most of them were poor",

Others attributed this to a lack of fruit and vegetables in the past:

"Because they didnt grow enough vegatables",
"We had more fruit".

Some answers appeared to be based on the findings of the 'skeleton in a box' activity:

"Because nearly every skeleton had a disease",

or based on personal reflection:

"because my teeth are cleaner that theres".

128 children (90.8%) felt they gained some ideas about how to eat more healthily and 138 (97.9%) felt they had gained some ideas about looking after their teeth. These themes were backed up with comments:

"You need to brush your teeth or you'll get plaque",
"I learnt you need to have vitamin C and iron",
"Don't have too much fizzy pop",
"Eating no fruit and veg is really bad",
"That in our time people have better toothbrushes".

Many children drew a tooth brush in the 'what I will remember most' section at the end of the questionnaire, particularly in the first year when we focussed on oral health.

4.2.1 Evaluation of Aims – Workshops for Schools

We believe we successfully engaged school children in human osteology; many of the children wrote about analysing their skeletons and working out the age and sex. Several drew pictures of the skeleton being laid out in the lab for 'What I will remember most'. We showed the effects of diet on teeth and jaws through the PowerPoint, archaeological skeletons and via the 'skeleton in a box' activity. Drawings in 'What I will remember most' included images of calculus, DISH, rickets and gout. There was less awareness (evidenced in the evaluation data) that similar diseases are seen in both archaeological and modern populations, but the answers to 'Do you think medieval people have healthier diets than us? Why?' suggested they were aware that different access to food – and especially sugar – has a huge impact on health. The children were also struck by the lack of dental hygiene in the past. Finally, we promoted the YAWYA project to teachers, encouraging them to book multiple YAWYA events by handing out merchandise at the end of the session. One school group clearly did some follow up work after the workshop, and sent us a series of beautiful thank you letters. Certainly one child thought the workshop would be memorable:

Figure 11.6. Illustrative feedback from a YAWYA osteology workshop for schools of Toby wearing a white suit.

"Everything was amazing. I will remember this in the future".

We had some unexpected outcomes. We raised the awareness of science (and the lab coats were a huge hit):

"It was like science",
"We wore lab coats, we saw the facts",
"how Toby looks in a white suit" (Figure 11.6).

One teacher observed that it was valuable to bring the children to a university for the workshop, as it could help raise their aspirations. At the start of the YAWYA project we did not anticipate these wider social outcomes.

5. Conclusions

This paper presents interim data: two further workshops for adults and four further schools workshops are planned for 2013. In addition, the YAWYA project has been recently awarded an extension, and four further schools workshops will run in 2014. Once these workshops have been completed, we will have more evaluation data, which will allow us to assess if the overwhelmingly positive responses reported here continue. We continue to modify the workshops in response to participants' and demonstrators' comments – for 2013 we intend to modify the section on the recording form 'what was the person called?' to 'What do you think the person might have been called? (Think about names that might have been common in the past)'; this change in language is hoped to convey that this aspect of the schools workshops is about imagination and reflection, rather than the scientific analysis of the skeleton.

The osteology workshops run as part of YAWYA are an extremely rewarding experience. We achieved all of our aims, and had some positive outcomes that we did not expect. We cannot assess long term reflection/knowledge gain, due to the scope of project, but the indications are that the workshops were memorable, and both adults and children gained knowledge about the impact diet can have on health. Overall the workshops seemed to have been a huge success:

> "...it was very fun and it was fun learning and I would want to come again".

Acknowledgements

We thank the Bradford workshop team: Julia Beaumont, Rhea Brettell, Emma Brown, Gemma Burton, Katie Clark, Alice de Jong, Rachel Haine, Rachel Holgate, Ceilidh Lerwick, and Jenni White, without whom the osteology workshops would not have been a success. John McIlwaine developed the first Bradford 'skeleton in a box' activity, and Kev Cale (Community Archaeology Ltd), helped modify 'skeleton in a box' to different scenarios prior to the YAWYA project. We also thank Maya Harrison, Fiona Blair, Gary Williamson and everyone else involved in YAWYA. Finally we thank the many workshop participants; we may have set out to teach them about osteology, but we all learnt from our engagement with them.

This project is supported by a Society Award from the Wellcome Trust, grant no. 092293.

Note

1. Key Stage 2 (or KS2) refers to the four years of school for children aged 7 to 11 years. This applies to England and Wales, but not Scotland or Northern Ireland.

References

Askew, S., Ross, C. 1988. *Boys Don't Cry: Boys and Sexism in Education*. Open University Press, Milton Keynes.

Brickley, M., Ives, R. 2008. *The Bioarchaeology of Metabolic Bone Disease*. Elsevier, Oxford.

Brooks, S.T., Suchey, J.M. 1990. Skeletal age determination based on the os pubis: A comparison of the Ascádi-Nemeskéri and Suchey-Brooks methods. *Human Evolution* 5, 227–238.

Brothwell, D.R. 1972. *Digging up Bones*. British Museum (Natural History), London.

Buikstra, J.E., Ubelaker, D.H. 1994. *Standards for Data Collection from Human Skeletal Remains*. Arkansas Archeological Survey, Fayetteville.

Fennema, E. 1996. Scholarship, gender and mathematics, in: Murphy, P., Gipps C. (Eds.), *Equity in the Classroom: Towards Effective Pedagogy for Girls and Boys*. Falmer Press, London, pp. 73–80.

Foster, H. 2008. *Evaluation Toolkit for Museums Practitioners*. East of England Museums Hub, Norwich.

Hein, G.E. 1998. The Constructivist Museum, in: Hein, G.E. (Ed.), *Learning in the Museum*. Routledge, London, pp. 14–40.

McCleery, I., Buckberry, J. 2010. *Rickets on the Increase?* http://www.leeds.ac.uk/yawya/news/news-rickets.html. Last accessed 1/2/2013.

Moore, W.J., Corbett, M.E. 1978. Dental caries experience in man: historical, anthropological and cultural diet-caries relationships, the English experience, in: Rowe, N.H. (Ed.), *Proceedings of a Symposium on Diet, Nutrition and Dental Caries*. University of Michigan School of Dentistry, Ann Arbor, pp. 3–19.

Museums, Libraries and Archives Council. 2008. *Inspiring Learning Checklists*. http://www.inspiringlearningforall.gov.uk/resources/theframework.html. Last accessed 31/1/2013.

Ogden, A.R. 2008. Advances in the paleopathology of teeth and jaws, in: Pinhasi, R., Mays, S. (Eds.), *Advances in Human Paleopathology*. Wiley, Chichester, pp. 283–307.

Payne, S. 2010. A child's gift to science. *British Archaeology* 112, 12–13.

Pearce, S.H.S., Cheetham, T.D. 2010. Diagnosis and management of vitamin D deficiency. *British Medical Journal* 340: b5664.

Roberts, C.A., Manchester, K. 1995. *The Archaeology of Disease*, 2nd edition. Sutton, Stroud.

Russell, T. 1994. The enquiring visitor – usable learning theory for museum contexts. *Journal of Education in Museums* 15, 19–21.

Smith, A. 1998. *Accelerated Learning in Practice*. Education Press, Stafford.

Younger, M., Warrington, M., Gray, J., Rudduck, J., McLellan, R., Bearne, E., Kershner, R., Bricheno, P. 2005. *Raising Boys' Achievement*. DFES Research report No 636. HMSO, Norwich.